Basal Ganglia: From Eminence to Function

Sachin

Table of contents

1. Introduction... 1

1.1. Spatial patterning of the embryonic telencephalon............................... 1

1.2. Development of GE- derived cells... 3

1.2.1. Generation of OLs... 4

1.2.2. Generation of INs.. 6

1.3. Regulation aspects of the chromatin remodelling complexes during 8
brain development..

1.4. Aims of this study and approaches... 10

2. Materials and methods.. 12

2.1. Experimental animals... 12

2.2. Fixation and tissue processing... 14

2.3. RNA sequencing (RNA-seq)... 14

2.3.1. RNA extraction 14

2.3.2. RNA sequencing.. 14

2.4. Fluorescent immunostaining... 15

2.5. *In situ* hybridization... 17

2.5.1. Probe generation by *in vitro* transcription (riboprobes)................... 17

2.5.2. Dig-labeling In situ hybridization.. 20

2.6. Image scanning and processing.. 21

2.6.1. Bright-field imaging... 21

2.6.2. Fluorescence microscopy.. 22

2.7. Schematic designs... 22

2.8. Quantifications and measurements... 22

2.9. Statistical analyses.. 23

3. Results... 24

3.1. Conditional deletion of BAF complex in late cortical progenitors alters 24
oligodendrogenesis in dcKO_hGFAP-Cre...

3.2. Expression of BAF155 and BAF170 in OL lineage................................ 26

3.3. Conditional deletion of BAF155 and BAF170 in OL lineage results in a global diminution of OL pool in developing forebrain... **29**

3.4. Reduction of proliferation capacity in BAF complex-deficient OPCs....... **35**

3.5. Expression of iOLs and mOLs markers are severely diminished in BAF155/BAF170-deficient subpallium... **37**

3.6. Cre fate mapping of cell lineages with Olig2+ history........................... **42**

3.7. Expression pattern of BAF155 and BAF170 in IN lineage.................... **46**

3.8. BAF complex ablation in the MGE progenitors leads to severe reduction of the GABAergic INs in the embryonic pallium **48**

3.9. Conditional lack of BAF complex affects the neuronal differentiation and migration.. **51**

3.10. Potential role of BAF complex in NSC fate choice............................ **53**

4. Discussion.. **55**

4.1. Expression of BAF155 and BAF170 in the OL lineage......................... **55**

4.2. BAF155 and BAF170 chromatin remodelling factor subunits are indispensable for OL development **56**

4.3. BAF complex controls the proliferative rate of the OPCs..................... **58**

4.4. Olig2 Cre mapping in glial versus neuronal derived progenitors........... **59**

4.5. Expression pattern of BAF155/170 subunits among IN lineage............. **60**

4.6. BAF complex controls the differentiation and migration of the MGE precursor cells.. **61**

4.7. Prospective role of BAF complex in cell fate decision......................... **63**

4.8. Conclusions and future perspectives... **66**

5. Summary.. **73**

6. References... **75**

1. Introduction

1.1.   Spatial patterning of the embryonic telencephalon

The telencephalon is the utmost complicated structure of the mammalian center nervous system (CNS) and displays enormous heterogeneity of cell lineages (Turrero Garcia and Harwell, 2017). During development, the anterior part of the telencephalon evaginates laterally forming a pair of telencephalic vesicles (Yun et al., 2001). These vesicles are subdivided into two structures: the dorsal pallium and the ventral subpallium (Wilson and Rubenstein, 2000; Yun et al., 2001) (Fig. 1A, B). The pallium represents mainly the presumptive cerebral cortex (Ctx), whereas the subpallium is partitioned into four anatomical parts; striatum (Str), pallidum (Pall), diagonal area (DG), and preoptic area (POA) (Medina and Abellán, 2012; Puelles et al., 2013; Puelles et al., 2016), which are the origin of the adult brain structures (Turrero Garcia and Harwell, 2017; Wilson and Rubenstein, 2000; Yun et al., 2001) (Fig. 1A, B). During early mouse brain development, the ventral neuroepithelium protruded into the lateral ventricles along the rostrocaudal axis of the forebrain, forming the three known ganglionic eminences (GE) and the developing POA (Turrero Garcia and Harwell, 2017). The GE is structured into three parts (Fig. 1B); the dorsal region is named as the lateral ganglionic eminence (LGE) and is the presumptive striatum. Whereas, the ventral region that is positioned medially with the regard to the LGE, identified as the medial ganglionic eminence (MGE); and is mainly the origin of the pallidal structures (globus pallidus, GP; ventral pallidus)(Turrero Garcia and Harwell, 2017). The medial ganglionic eminences (MGE) and lateral (LGE) ganglionic eminences mainly consider the presumptive striatal and pallidal structures of the basal ganglia. Each domains of GE (L/M/CGE) are further subdivided into three strata: the ventricular zone (VZ), the subventricular zone (SVZ) harbouring progenitors, and the mantle zone (MZ) which comprises primarily differentiated neurons and others (Chen et al., 2017; Flames et al., 2007; Petryniak et al., 2007). Anatomically, CGE is constructed by the fusion of the LGE and MGE (Turrero Garcia and Harwell, 2017; Watson et al., 2012).

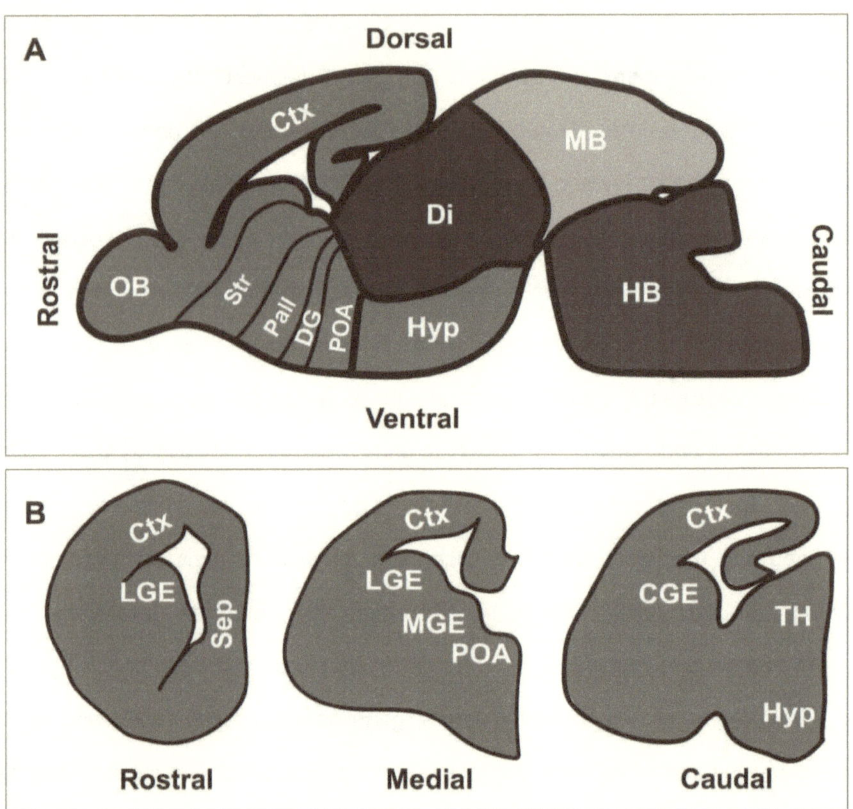

Figure 1. Major subdivisions of the mammalian telencephalon. (A) Sagittal section represents the major structures of the developing murine brain. (B) Coronal sections depict the different levels of the telencephalon that have employed over this study. Abbreviations: OB, olfactory bulb; Ctx, Cortex; Str, striatum; Pall, pallidum; DG, diagonal area; POA, preoptic area; Hyp, hypothalamus; Di, diencephalon; MB, midbrain; HB, hindbrain; LGE, lateral ganglionic eminence; Sep, septum; MGE, medial ganglionic eminence; CGE, caudal ganglionic eminence; TH, Thalamus.

1.2. Development of GE- derived cells

The neural stem cells (NSCs) characterize by two cardinal properties: cell-renewal capacity and the ability to differentiate into other cell types such as committed progenitors and post-mitotic cells. The NSCs give rise to diverse neuronal and non-neuronal cell lineages within the telencephalon (Adams and Morshead, 2018; Bond et al., 2015; Ma et al., 2009; Xu et al., 2017) (Fig. 2A, B).

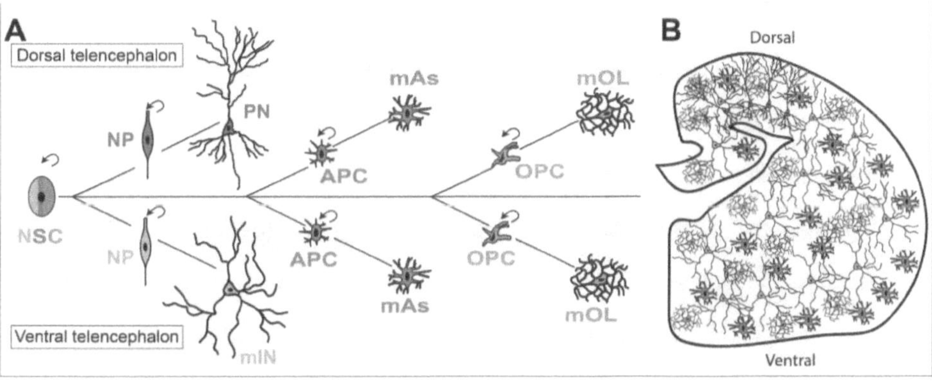

Figure 2. Schematic diagram illustrates heterogeneous cell lineages in the mammalian forebrain. (A) Schematic illustration depicts the diverse cell lines that could be produced from the multi-potent NSCs in the mammalian forebrain. (B) Cartoon of coronal brain section of telencephalon illustrates the spatial distribution of the different cell lineages within the murine telencephalon. Colour refers to the diverse cell lines; the neuronal lineages showed in pink colour, oligodendrocytic lineages in light blue colour, and astrocytic lineages in green colour. Abbreviations: NSC, neural stem cell; NP, neural progenitor; PN, pyramidal neuron; mIN: mature Interneuron; APC, astrocyte precursor cell; mAs, mature astrocyte; OPC, oligodendrocyte precursor cell; mOL, mature oligodendrocyte; Ctx, cortex; LGE, lateral ganglionic eminence, MGE: medial ganglionic eminence, POA: preoptic area.

1.2.1. Generation of OLs

Among the non-neuronal cell populations, OLs are unique myelin-producing cells within the central nervous system (CNS). Noteworthy, myelin plays a pivotal role in the saltatory transmission of action potentials along neuronal axons. Furthermore, OLs are recruited to supply neurons with trophic support (Bradl and Lassmann, 2010; Nave, 2010). Hence, OLs are principal cells in CNS not only for the mending of neural transduction, however for the neuronal maintenance, as evinced by the neurodegenerative diseases studies that linked to OLs demyelination. The myelinated OLs development is a multistep process starting with the specification of the oligodendrocytes precursors (OPCs) and ending with the exit of the cell cycle, triggering the differentiation and myelination programs (Bergles and Richardson, 2015; Richardson et al., 2006). Cell-tracing studies revealed that OLs are generated from the ventral and dorsal telencephalon. The first OL wave is generated at 12.5 from Nkx2.1-derived progenitors in MGE, followed by OL set at E15.5 from the GSX2 in LGE and CGE. The latest OL group of the cortical Emx1-expressing progenitors at birth and ultimately represent the main pool of the adult OLs (Kessaris et al., 2006; Naruse et al., 2017; Newville et al., 2017) (Fig. 3A). Additionally, the cell-fate patterning necessitates the expression of cascade genes, which precisely activate a lineage-specific program. Numerous studies have revealed diverse transcriptional modulators of the OL production (Bischof et al., 2015; Goldman and Kuypers, 2015; Matsumoto et al., 2016; Yu et al., 2013) (Fig. 3B). For instance, the basic helix-loop-helix (bHLH) Olig2 protein acts as a master transcription factor (TF) of the OL lineage by governing OPC specification, differentiation, and myelination. Moreover, the stepwise specification of OPCs is driven by a defined set of TFs, including Olig1,

PDGFR (Platelet-Derived Growth Factor Receptor, Alpha), Sox10 (SRY-box10), Zfp488 (zinc finger protein 488) and Zfp191 (zinc finger protein 191) (Bischof et al., 2015; Yu et al., 2013).

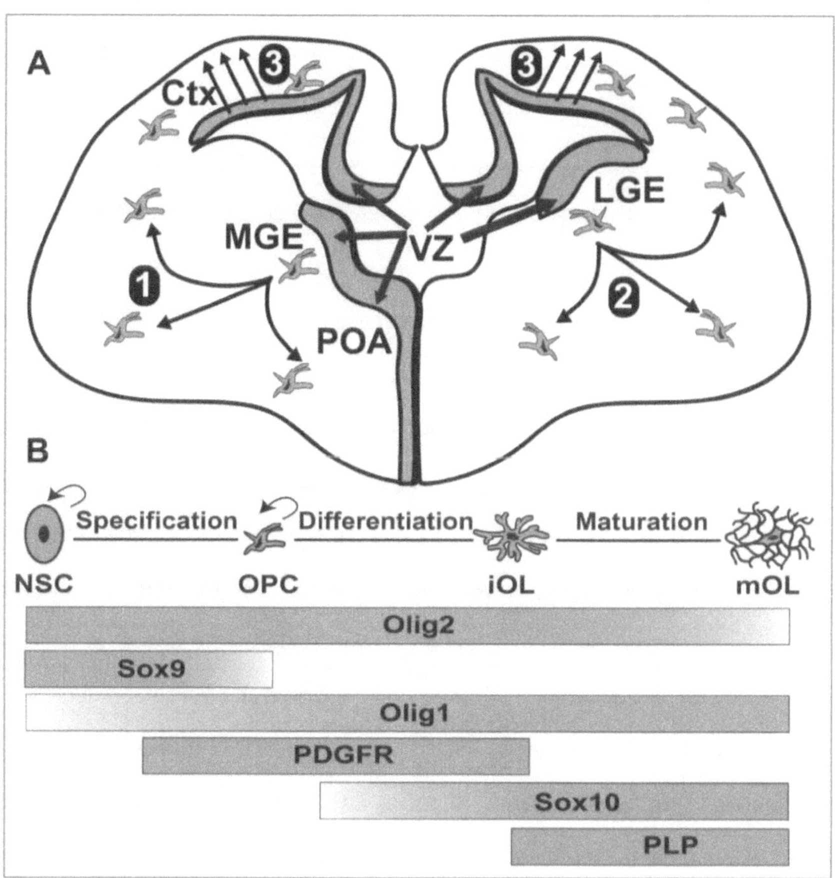

Figure 3. Schematic representation of oligodendrogenesis in the murine forebrain. (A) Origin of the OPCs and their migratory pathways to the cortex. Three OPCs waves of different birthdates are: (1) From Nkx2.1 expressing progenitors in POA and MGE at E12.5, (2) from GSX2 expressing progenitors in LGE and CGE at E14.5, (3) from Emx1 expressing progenitors in the cerebral cortex after birth. (B) Schematic illustration of diverse phases of OL progression from NSC to gliablast, OPC toward iOL and mOL and indicated OL lineage markers. Abbreviations: Ctx,

cortex; MGE, medial ganglionic eminence; POA, preoptic area; LGE, lateral ganglionic eminence.

1.2.2. Generation of INs

GE is the main origin of a vast number of progenitors, which form the presumptive cortical cells, such as oligodendrocytes (OLs), interneuron (IN) subtypes, and astrocytes (Turrero Garcia and Harwell, 2017). Interestingly, the subpallial NSCs generate also INs (Fig. 4 A, B). Shortly after specification of the IN precursors, they undergo the differentiation program along with the various stereotyped migratory streams toward the cortex. IN subtypes slowly express their specific molecular markers at the onset of their maturation process (Que et al., 2019; Wamsley and Fishell, 2017). Gene expression profiling studies have identified the transcriptional signatures that orchestrate INs development (Fig. 4 A, B). Intriguingly, the heterogeneity of the INs subtypes is basically associated with the transcriptional topography of the GE (Kessaris et al., 2014) (Fig. 4 A, B). For example, the spatial identity of the MGE and POA is bestowed by Nkx2.1 TF that specified the MGE progenitors (Flames et al., 2007; Flandin et al., 2010; García-López et al., 2008; Puelles et al., 2000; Sussel et al., 1999) and its expression prevails in subgroups of postmitotic neurons (Flandin et al., 2010; Marın et al., 2000; Sussel et al., 1999). Furthermore, a cascade of genes such as Dlx family (Dlx1, Dlx2, Dlx5, Dlx6) interplay with the transcriptional elements; for instance Nkx2.1, Lhx6 and Sox6, regulating the development of the MGE-derived INs commencing by their specification ending with maturation and migration steps (Kessaris et al., 2014; Touzot et al., 2016) (Fig. 4 A). Moreover, the MGE can generate other kinds of neurons; including pallidal projection neurons, striatal interneurons and striatal cholinergic neurons (Chen et al., 2017; Flandin et al., 2010; Marın et al., 2000). On the other hand, the peculiarity of LGE and CGE progenitors is determined by the homeobox TF GSX2. Specifically, the expression of the nuclear receptor COUP-TFII (known as Nr2f2) enhances the CGE progenitors to migrate toward the cerebral cortex (Kanatani et al., 2008; Touzot et al., 2016; Tripodi et al., 2004). Based on the spatial and temporal transcriptional landscape of the GEs, diverse neuronal subtypes have been engendered. MGE and POA produce nearly all INs in the SST+ and PV+ groups (Gelman et al., 2011; Kelsom and Lu, 2013; Lim et al., 2018; Touzot et al., 2016), whereas, all VIP+ and CCK+ classes are derived from the CGE (Lim et al., 2018; Niquille et al., 2018). In addition to the POA that generates virtually the NPY+ INs and Neurogliaform cells (Gelman et al., 2009; Lim et al., 2018; Niquille et al., 2018). Previous studies have been broadly characterized the assortment of INs on diverse basis; for instance, morphology, transcriptional profiles, electrophysiological activities, neural synaptic configurations (Ascoli et al., 2008; Lunden et al., 2019; Staiger et al., 2015). Indeed, the neuronal heterogeneity along the cerebral cortex reveals a crucial function in the neural connectivity of the developing CNS. Of greatest significance is that cortical INs role in modulating the excitatory and inhibitory circuitries (Hensch, 2005; Kelsom and Lu, 2013). Importantly, cortical INs are defined by a cardinal property, which is the releasing of the inhibitory neurotransmitters termed gamma-aminobutyric acid (GABA) (Kelsom and Lu, 2013; Somogyi et al., 1983). Thus, GABAergic INs conduce the network connectivity among the excitatory pyramidal neurons by suppressing the activity of the desired hyperpolarized neurons, and subsequently inhibiting the possibility of neuronal firing (Lunden et al., 2019; Marín, 2012; Selten et al., 2018).

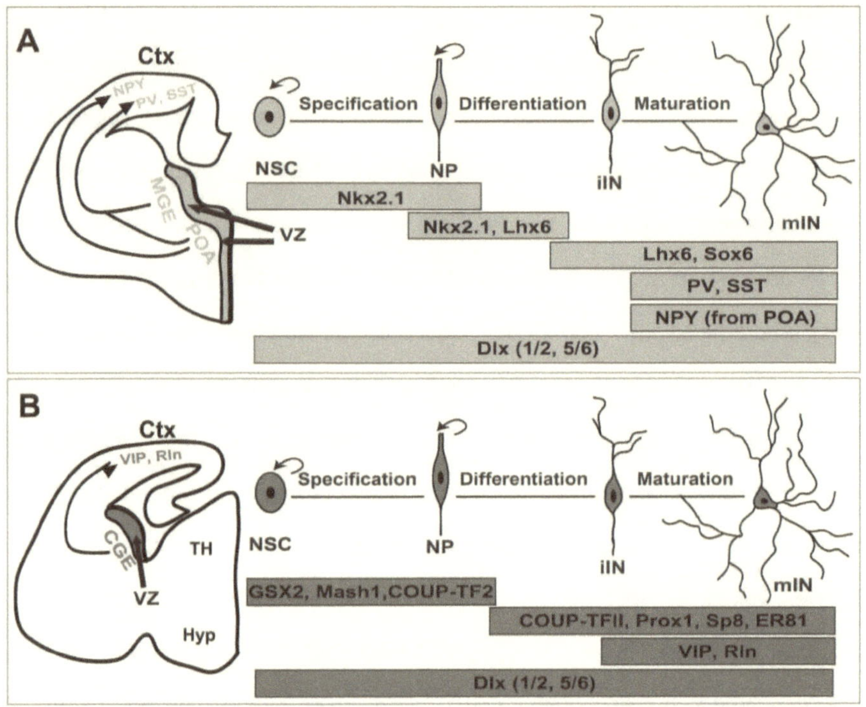

Figure 4. (A, B) Diagrammatic illustrations show the cortical INs development. (A) Origin of MGE derived INs and their migratory pathways to the cortex. Transcriptional markers contribute in the development of the MGE born INs. (B) Origin of CGE derived INs and their migratory pathways to the cortex. Transcriptional factors contribute in the development of CGE born INs Abbreviations: TH, thalamus; Hyp, Hypothalamus; MGE, medial ganglionic eminence; CGE, caudal ganglionic eminence; POA, preoptic area; NSC, neural stem cell; NP, neural progenitor; iIN, immature interneuron; mIN, mature interneuron.

1.3. Regulation aspects of the chromatin remodelling complexes during brain development

During mammalian brain development, various integrated epigenetic mechanisms interplayed to permit the accessibility of TF to the chromatin and the subsequent gene expressions; involving DNA methylation (Alfert et al., 2019; Jin and Liu, 2018;

Moore et al., 2013), histone modifiers (Alfert et al., 2019; Audia and Campbell, 2016; Bannister and Kouzarides, 2011) and chromatin re-modellers (Alfert et al., 2019). One paramount example of the vertebrate chromatin modifiers is the mammalian ATP- dependent BAF complex, which utilizes the energy produced from the ATP hydrolysis for alteration of the histones and nucleosomes (Hargreaves and Crabtree, 2011; Yoo and Crabtree, 2009). BAF complex structurally composed of assembly of up to 15 subunits; that encompass core and peripheral subunits (Kadoch and Crabtree, 2015; Narayanan et al., 2015). Among them, either Brg1/Smarca4 or Brm/Smarca2 with their ATPase activity, together with the core units and BAF155 and BAF170 that act as the scaffolding units (Narayanan et al., 2015). Peripheral elements associate to the core subunits revealing a broad variability, which displays several potential rearrangements of the BAF assembly subunits (Mashtalir et al., 2018; Narayanan and Tuoc, 2014). *In vitro* studies demonstrated that Brg1/ Brm1 alone is able to remodel the nucleosome; the additions of other core subunits BAF47, BAF155 and BAF170 enhances remodelling activity (Alfert et al., 2019; Narayanan et al., 2015; Phelan et al., 1999) (Fig.4). The structure and dynamic variations of the BAF complex is assumed to provide the BAF complex the capability to control gene expression of various cell lines (Ho and Crabtree, 2010; Sokpor et al., 2017). For instance, BAF45a and BAF53a are expressed in the neuronal progenitors express, however, they are substituted by BAF45b and BAF53b in the differentiated neurons (Ho and Crabtree, 2010; Yoo and Crabtree, 2009). Another example, the pluripotent embryonic stem cells are linked to the expression of some BAF complex-subunits involving Brg1, BAF155, BAF250a, and BAF60a/b, but not other subunits including Brm, BAF170, BAF60c, and BAF250b (Kidder et al., 2009; Sokpor et al., 2017). Combined BAF complex subunits interplay with various transcription regulators and signalling pathways, and thus are indispensable for genome targeting and proper gene functional specificity (Ho and Crabtree, 2010; Narayanan et al., 2015; Narayanan and Tuoc, 2014). Various subunits of the BAF complex assist to study its function with respect to different cell lines over the brain development (Nguyen et al., 2018; Tuoc et al., 2017). Previous phenotype analyses displayed the conditional knockout mouse model, which revealed a complete loss of the entire BAF complex. This can be resulted by removal of the scaffolding BAF subunits (BAF155 and BAF 170), resulting in unbinding of the BAF subunits and subsequently ultimate degradation by the ubiquitination system (Narayanan et al., 2015).

1.4.   Aims of this study and approaches

Mediation of various Cre driver lines creating conditional BAF155_BAF170 double knockouts is the ideal model to study involvements of BAF complex in brain development. Our former works have comprehensively investigated the role of BAF complex in regulating the dorsal telencephalon (Narayanan et al., 2015; Nguyen et al., 2018); for instance, the phenotype of the forebrain specific dcKO_FoxG1-Cre mice displayed complete loss of the forebrain (Narayanan et al., 2015). Whereas, lack of the cortical BAF complex interceded by Emx1-Cre-mediated demonstrated impaired underdeveloped pallium (Narayanan et al., 2015). Additionally, our recent studies manifested that hGFAP-Cre mediated conditional loss of BAF complex resulted in upregulation of the proliferative pool of the cortical NSCs (Nguyen et al., 2018). Although our previous studies have shed light on the significance of the BAF complex in controlling the transcriptional program of the corticogenesis, insights into the BAF complex trajectories that orchestrate cell fate of the subpallial progenitors are still over-simplistic. Thus, this study aims to investigate the possible implication of the chromatin modifiers BAF complex in the differentiation of ventral NSCs that produce OLs and INs lineages. Previous works revealed that Brg1-dependent chromatin remodelling complex is recruited by the Olig2 for oligodendroglial progression and differentiation (Bischof et al., 2015). Furthermore, Yu et al. (2013) reported that the conditional loss of Brg1 interceded by Olig1-Cre resulted in a complete block of OL differentiation. Nevertheless, the involvement of the BAF complex that govern OPCs proliferation is still elusive. Thus, the first aim of this work is to elucidate the function of BAF complex in governing the proliferation of the OPCs. On the other hand, current works have cast light on studying the pivotal role of the AT-rich-interactive-domain-containing protein 1b (Arid1b) - dependent chromatin remodelling complex in brain development of the cortical INs (Celen et al., 2017; Jung et al., 2017). Remarkably, Arid1b-cKO and heterozygote mice showed normal development of the different cell lines in the forebrain, except the GABAergic INs that were strongly decrease in number, henceforth, imbalance of the cortical excitatory and inhibitory circuits (Jung et al., 2017). Nonetheless, we are still far from understanding the function of the BAF complex in governing the proliferation, differentiation and migration of the different types of the pallial INs. Therefore, the second objective of this study is to discern the prospective implication of BAF complex in regulating cell fate decision and subsequently the development of IN lineage over the embryogenesis. Thus, we conducted the conditional double-knockout approach (dcKO) in the mouse, where BAF155 and BAF170 deletion is limited to the ventral subpallial progenitors (origin of OL line as well as IN line).

2. Materials and methods

2.1. Experimental animals

Homozygous floxed BAF155 (Choi et al., 2012), Homozygous floxed BAF170 (Cong Tuoc et al., 2013), hGFAP-Cre (Zhuo et al., 2001), Olig2-Cre (Schüller et al., 2008), tdTomato (Ai9) (Madisen et al., 2010) mouse lines were provided by the Jackson Laboratory, Bar Harbor, USA and maintained in a C57BL6/J background. In most of the experiments, we employed hGFAP-Cre and Olig2-Cre to generate pallium-specific BAF 155/170 dcKO _ hGFAP_Cre and subpallium-specific BAF 155/170 dcKO _Olig2_Cre offspring, respectively (Fig. 5A, B).

Moreover, heterozygous Olig2-Cre mouse line was bred with the homozygous floxed tdTomato mice (Ai9) mice. The resulting offspring (BAF 155/170 heterozygote_Olig2 Cre _tdTomato) and (BAF 155/170 dcKO_Olig2 Cre _tdTomato) was conducted for cell lineage tracing studies (Fig. 5C). Husbandry of mice was performed based on the guidelines approved by the University Medical Center Göttingen (Göttingen, Germany). The practical manipulations were carried out following German laws on animal research protection (Zintzsch, 2013).

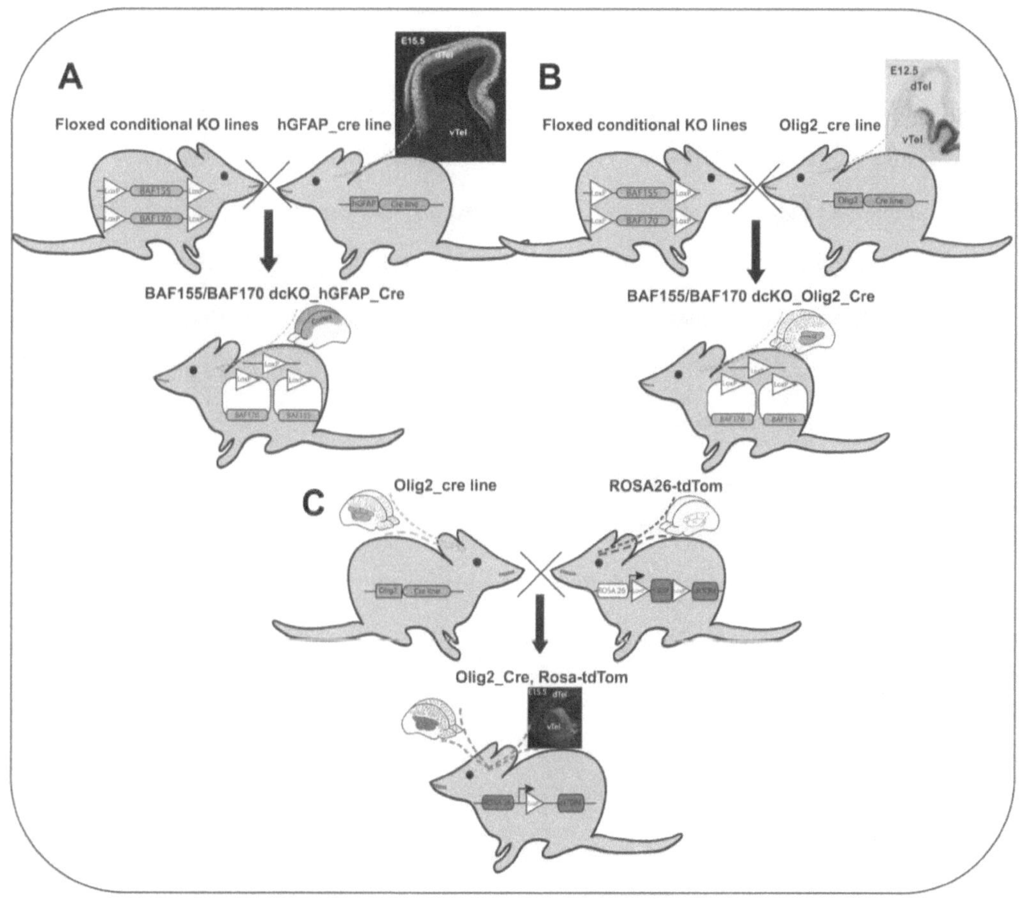

Figure 5. Schematic illustrations depict the generation of the transgenic mouse lines employing different Cre lines. (A) Cartoon depicting the creation of the BAF155/BAF170 dcKO mouse line conducting hGFAP Cre that is actively spread in the pallium around E15.5. (B) Schema illustrating the generation of the BAF155/BAF170 dcKO mouse line using Olig 2 Cre that is actively expressed in the subpallium around E12.5 in the GE. (C) Diagram explaining the creation of the BAF 155/170 dcKO_Olig2 Cre _tdTomato conducting Olig2 Cre to label all Olig2+ progeny cells. Abbreviations: dTel, dorsal telencephalon; vTel, ventral telencephalon; tdTom, tdTomato.

2.2. Fixation and tissue processing

Pregnant animals with the target age of embryos (E13.5 - E18.5) or postnatal animals (P3) were sacrificed. Tail biopsies were collected from each specimen (embryo or adult) for DNA extraction and genotyping by PCR. Embryonic and postnatal brains were dissected in DEPC-treated H_2O (on ice), subjected to 4% PFA for fixation for 2 h (E15.5), 2.5 h (E17.5), 2.5 h (E18.5) and 3.5 h (P3). Afterwards, fixed brains were immersed in 25% sucrose prepared in 0.01 M PBS overnight at 4°C for dehydration. Then, the specimens were embedded in blocks and preserved in Tissue-Tek, cryo-sectioned at 10 µm thickness (E13.5 - E 17.5) or 16 µm thickness (18.5 and P3). Then, the slides were stored at -20°C until their usage.

2.3.   RNA sequencing (RNA-seq)

RNA sequencing was accomplished to compare the transcriptomic profiling of the cortices (P3) in WT and dcKO mice.

2.3.1. RNA extraction

The total RNA was extracted from P3 brains employing RNA purification kit (RNeasy Plus Mini, Qiagen) based on the manufacturer's instruction. Purified RNAs were stored at -80°C until further investigations.

2.3.2. RNA sequencing

RNA-seq, including cDNA library preparation, sequence process and assembly of the transcriptomic data were performed in collaboration with Prof. André Fischer's laboratory. cDNA libraries were produced using the Library Prep Kit (TruSeq RNA Library Prep Kit v2, Illumina) according to the attached protocol. FASTQ conversions and base calling were accomplished utilising Illumina templates as demonstrated earlier (Halder et al., 2016). The collection of the transcriptome was mapped based on mm10 mouse reference genome adopting STAR aligner v2.3.0 as formerly designed (Djebali et al., 2012) and ensuing differential gene expression investigations were achieved using DESeq2, Bioconductor as designated before (Love et al., 2014). Gene enrichment set analysis (GSEA) were analysed using WebGestalt (WEB-based Gene Set Analysis Toolkit) http://www.webgestalt.org/ as described in (Wang et al., 2013).

2.4. Fluorescent immunostaining

The immunohistochemical approach was accomplished employing the combinatory method of peroxidase anti-peroxidase (PAP)- and avidin-biotin-peroxidase complex (ABC), as formerly described (Davidoff and Schulze, 1990).

After washing the sections for 1 h in 0.01 M PBS (2 x15 min), the sections were kept overnight at 4 °C in primary antibodies after blocking with normal sera of the suitable species. To visualize the nucleoside analogues like 5-iodo-2'-deoxyuridine (IdU), a pre-treatment with 2N HCl at 37 °C for 30 min was applied prior to the incubation with the respective primary antibodies. Next, tissue was rinsed with PBS and then immunostained for 2 h with the respective secondary antibodies that were prepared in the blocking solution (Alexa Fluor, 1:400; Invitrogen). Then, nuclei were stained with 4'6-diamidino-2-phenylindole (DAPI) (1:1000). Double and triple immuno-staining was carried out sequentially (one antigen followed by the other one). Lastly, brain tissue was mounted using Aqua-Poly/Mount (Polyscience).

Antibodies

The applied primary antibodies, polyclonal (pAb) and monoclonal (mAb) primary antibodies, dilutions and were applied as follows:

Antibody	Host	Dilution (1:)	Company
BAF155	Rabbit	250	Santa Cruz
BAF155	Mouse	200	Santa Cruz
BAF170	Rabbit	100	Bethyl
BAF170	Mouse	200	Santa Cruz
Olig2	Rabbit	500	Millipore
PDGFR	Rat	200	BD
Sox9	Rabbit	500	Millipore
Sox10	Guinea pig	400	Michael Wegner
Casp3	Rabbit	200	cell signalling
Ki67	Rabbit	50	Novocastra
Ki67	Mouse	100	Novocastra
IdU	Mouse	50	Becton Dickinson
Nkx2.1	Mouse	50	Progen
GABA	Rabbit	200	Sigma
Sox6	Rabbit	500	Abcam
BLBP	Rabbit	200	Chemicon/Millipore

Table 1. Primary antibodies applied for IHC.

Secondary antibodies applied were Alexa 488-, Alexa 568-, Alexa 594- and Alexa 647-conjugated IgG (different species, 1:400).

2.5. *In situ* hybridization

Chromogenic *in situ* hybridization (ISH) experiments were conducted on 10-µm slices from embryonic brains (E15.5) and 16-µm sections from developing brains (E18.5). The embryonic brains were fixed in 4% paraformaldehyde and cryopreserved. The detection of the RNA transcripts of different RNA probes (riboprobes) was visualized via staining of the chromogen Digoxigenin (DIG)-marked specific riboprobes. The protocol was updated based on Wagener et al. (2010). To reduce the possibilities of RNA debasement, the experiment was manipulated in RNAse free circumstances, in addition to the conceivable processes were applied to evacuate RNA-digesting RNAses. Therefore, the used reagents were treated with DEPC and autoclaved afterwards. Additionally, the glassware was sterilized for at least 6h heated in 190°C oven (Memmert, 400).

2.5.1. Probe generation by *in vitro* transcription (riboprobes)

In Situ Hybridization (ISH) for different OL markers such as Olig2, Olig1, Sox10 and PLP and neuronal indicators involving Olig1, Olig2, Sox10, PLP, Dlx2, Nkx2.1, and Lhx6 were carried out using single-strand antisense and sense digoxigenin (DIG)-labelled cRNA probes. Probes were generated straight from the plasmids including the specific genes.

Gene	Name	Function	Ensembl ID
Olig1	Oligodendrocyte transcription factor 1	Stimulates maturation and development of OLs in the brain	ENSG00000184221
Olig2	Oligodendrocyte Transcription Factor 2	Transcriptional regulator of ventral NSCs fate, promotes OL development and inhibit Nkx2.1 born IN development.	ENSG00000205927
Sox10	SRY-Box Transcription Factor 10	Enhances expression of myelin genes, during OL maturation.	ENSG00000100146
PLP	Proteolipid Protein 1	Predominant constitution of myelin, maintenance of myelin sheaths, OL development.	ENSG00000123560
Dlx2	Distal-Less Homeobox 2	Master activator for IN development, regulate terminal differentiation of INs.	ENSG00000115844
Dlx1	Distal-Less Homeobox 1	Play crucial role in vTel, regulate terminal differentiation of INs.	ENSG00000144355
Nkx2.1	NK2 Homeobox 1	Promote MGE born INs development.	ENSG00000136352
Lhx6	LIM Homeobox 6	Cortical INs specification and migration from the ventral telencephalon to the cortex.	ENSG00000106852

Table 2. Genes designated for ISH.

Plasmids employed in this project

The plasmids were amplified in the competent bacteria *(E. coli)* and afterwards purified employing midi and maxi Qiagen Kit. Purified plasmid cDNA was linearized by the incubation with the adequate restriction enzymes (table3).

Gene	Restriction enzymes	Polymerase enzymes
Olig1	SacI	T3
Olig2	EcoRI	T7
Sox10	BamHI	T7
PLP	BamHI	T3
Dlx2	EcoRI	T3
Nkx2.1	XbaI	T3
Lhx6	NotI	T3

Table 3. Synthesized probes for *in situ* hybridizations.

To check the linearization efficiency, the digestion elution was loaded, after the enzymatic reaction was ended, on 1.5% agarose gel (1.5% agarose in TBE buffer). Then, the linearized plasmids were purified using NucleoSpin^R Gel and PCR Clean-up, Maherey-Nagel kit, Cat. 740609.240C).

To synthesize the anti-sense mRNA probes, the linearized purified plasmids were transcribed *in vitro* employing Digoxygenin-labelled (DIG) UTP´s. The reaction comprises 2 µl of DIG-Nucleotide mix (Boehringer), 2 µl of 10X transcription buffer

(Roche), 1 µl of RNA polymerase (Roche), 1 µg of the purified DNA, 1 µl of RNAse inhibitor (Promega), and 13 µl of DEPC-treated H2O. Then, the mixture was maintained at 37°C for 2 h incubation. To assure the complete degradation of the excessive template DNA, 2 µl/reaction DNAse was supplemented to the mixture at 37 °C for 10 min. Finally, 2 µl DEPC treated dH_2O was added to the reaction mix to stop the transcription reactions. Once transcribed, the synthesized probes were obtained by adding the pre-cooled (-20 °C) mixture containing 4M Lithium Chloride and 100% ethanol to the *in vitro* transcription products and incubated for at least 30 min. at -80 °C. Afterwards, the tubes were centrifuged for 45 min. at 13,000 rpm at 4 °C; the precipitated probes were rinsed in 70% Ethanol and then allowed to dry. The RNA pellets were subsequently dissolved in 50 µl DEPC treated H_2O and stored at – 80 °C until use.

2.5.2. Dig-labeling *in situ* hybridization

On ice, tissue was subjected to post-fixation using 4% PFA (20 min) and washed in PBS (2 x 10 min). Slides were incubated in 1% H_2O_2 in methanol to reduce the endogenous peroxidase reaction for 15 min and then transferred for washing in PBS twice 2 min each time. Next, brain tissues were deproteinized by applying 0.2 N HCl for 8 min, then were rinsed in PBS for 2 min and thereafter the permeability for probe penetration was employed for 3 min using proteinase K (20 µg/ml) dissolved in 0.05 M TBE pH 7.5. Slices were washed (PBS, 1x 5 min), following 20 min treatment with cold 4% PFA for 20 min, tissue was subjected to 0.2% acetic anhydride dissolved in triethanolamine hydrochloride (0.1 M) for background reduction. Sections were rinsed in PBS and once in 2 x saline sodium citrate (SSC) buffers for 5 min.

Afterwards, tissue was subjected to series of dehydration steps employing alcohols in the subsequent order: for 20 sec in 30% ethanol (20 sec), 20 sec in 50% ethanol

(20 sec), 1 min in 70% ethanol, 20 sec in 80% ethanol, 20 sec in 95% ethanol and 2 times 20 sec in 100% ethanol. slides were then pre-hybridized for minimum 1 h at 55°C in hybridization buffer (4XSSC, 5% dextran sulfate, 50% formamide (deionized) (v/v),250 µg/ml HSS-DNA, 100 µg/ml tRNA, 1X Denhardt's solution) and hybridized once more in hybridization buffer overnight with the DIG-labelled with the desired probes at 55°C. Afterwards, brain sections were subjected to stringency rinses as follow: 1 min in 5×SSC at 65 °C, 30 min in 2×SSC/50% formamide at 65 °C, 30 min in 1×SSC/50% formamide at 65 °C, 30 min in 0.1×SSC at 65 °C. Later, the tissues were washed 3 times (2 min per each) with TBS at 30 °C and then were incubated with 1% blocking buffer dissolved in TBS (pH 7.5) for 30 min at room temperature. Later, tissue was incubated overnight at 4°C in a humidified chamber with anti-DIG-alkaline phosphatase (AP) antibody 1:500 (Roche,11093274910) in blocking solution. Next day, the sections were subjected to many wash steps (TBS pH 7.5, 2x 10 min; reaction buffer 100 mM Tris-HCl, 100 mM NaCl pH 9.5 and 50 mM MgCl2, 1 x 10min), the different probes were detected by the NBT/BCIP colour reaction and then the slides were coverslipped as using Kaiser's glycerol gelatine medium. As negative control experiments, certain slides were also applied to the sense probes, and then developed in parallel.

2.6.   Image scanning and processing

2.6.1. Bright-field imaging

The bright-field scanning was acquired using 10x and 25x objective (LCI Plan-Apochromat, Zeiss) of an Axio Imager M2 (Zeiss) Apotome. The Neurolucida software (MBF Bioscience, version 11) is the software that organizes this microsope and CCD camera (Retiga 2000R, Qimaging) is the camera used to capture the bright-field images. Images were captured as image stacks and the final bright-field images are exported and visualized as minimum intensity projections.

2.6.2. Fluorescence microscopy

Fluorescent images were scanned by means of the same microscope, nevertheless with a structured illumination module (ApoTome, Zeiss) operated by the Neurolucida software. Furthermore, the spectral confocal microscope (TCS SP5, Leica) was used for high scanning resolution. These confocal micrographs were recorded as either single image or image stacks using 40x objective (40.0x1.25 OIL UV, Leica) and 63x objective (63×/1.40-0.60 OIL, Leica), Leica software (Las AF, Leica) is operating software of this microscope. The final confocal stacks images were merged and shown as maximum intensity projections. Suitable image improvements; for instance, colour balance levels were modified using Adobe Photoshop (version CS6) and/ or FIJI. Figures were arranged in Adobe Photoshop CS6.

Confocal microscopies were achieved at the Facility for Innovative Light Microscopy (FILM) at the Max Planck Institute for Biophysical Chemistry, Göttingen.

2.7. Schematic designs

Adobe Illustrator CS6 was used to draw the diagrammatic figures: 1 (A and B), 2, 3, 5, 7(A), 16(A), 20(A-C) and 21.

2.8. Quantifications and measurements

Cell counting was manipulated employing different professional software such as Neurolucida software (MBF Bioscience), Leica and image Fiji.

The sections among the experimental groups (n= 2-8) according to their availability were selected as they were comparable in the rostrocaudal axis. Concerning cell quantifications with fluorescence signal sections, positive cells were recorded in the targeted area with identically sized frames of the particular regions (str, ctx and GE).

Concerning cell counts with the chromogenic signal, cells expressing desired probes were counted directly by Neurolucida software within desired zones which represented by the whole brain hemispheres (half brain). Fluorescence intensity signals were measured using Image Fiji software, as formerly demonstrated (Narayanan et al., 2015; Tuoc and Stoykova, 2008). Heat map analyses have been conducted by means of image Fiji.

2.9.   Statistical analyses

Datasets and statistical analyses were processed by means of graph prism (version 5) for calculation of the arithmetic average and the standard deviation, as well as the checker of the significance. The statistical variances among groups were evaluated adopting two-tailed unpaired Student t-test. The variances between groups were statistically significant at p-value less than 0.05. The significant levels are as follow; *p ≤ 0.05, **p ≤ 0.01, ***p ≤ 0.001, ****p ≤ 0.0001. All graphs shown in this book were plotted by GraphPad Prism software (version 5).

3. Results

3.1. Conditional deletion of BAF complex in late cortical progenitors alters oligodendrogenesis in dcKO_hGFAP-Cre

Our previous study pointed out that the deletion of BAF complex in the late corticogenesis led to impaired neurogenesis in parallel with the increased pool of NSCs in BAF155/BAF170 dcKO_hGFAP cortex at prenatal stages (Nguyen et al., 2018). To examine a possibility that increased pool of NSCs in dcKO_hGFAP cortex (Fig. 6A) generate more glial cells, a comparative transcriptomic analysis was performed for the cortical cells that were pre-isolated from the cortex of postnatal mice at P3. Unexpectedly, gene enrichment set analysis (GSEA) showed downregulation of the genes, which are involved in oligodendroglial development (Olig2, Olig1, PDGFR, Mbp, and Mag, among others) in the dcKO pallium relative to the control counterparts in (Fig. 6B, C). To assess the possible function of the BAF complex in the regulation of oligodendrogenesis in the postnatal mouse brain, mediating the Cre-dependent conditionally knockout of BAF155 and BAF 170 in cortical neural stem cells. Therefore, the BAF155/170 floxed (BAF155$^{fl/fl}$ /BAF170$^{fl/fl}$) mice are mated with the hGFAP-Cre mice line to create dcKO mutants. At E15.5, we previously demonstrated that the hGFAP-Cre is strongly active in the cerebral cortex of the mouse brain (Nguyen et al., 2018). We further substantiated our findings adopting the immunostaining with PDGFR antibody that labels the OPCs in the adult cortex. In the absence of BAF155/170, the number of PDGFR + OPCs in the mutant brain was depleted compared to the control as depicted in Fig. 6D. The findings are consistent with the results obtained from the transcriptomic analysis, in which, the quantitative comparisons of the dcKO cortical PDGFR + OPCs that are normalized to the DAPI+ cells were significantly reduced compared to that of the control brain (Fig.

6D`). Collectively, these outputs reveal that absence of BAF complex during corticogenesis suppresses the induction of the genes that are linked to OL production, which point to the crucial role of BAF complex in promoting the development of oligodendrogenesis.

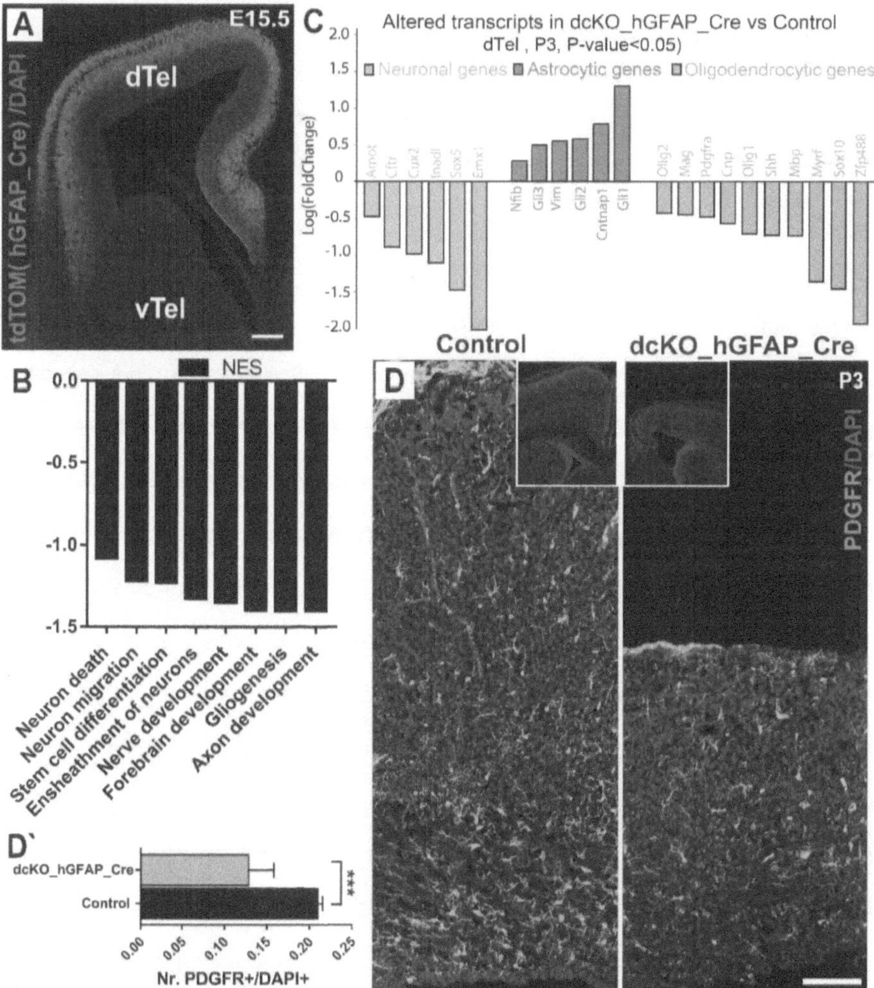

Figure 6. Cortex-specific loss of BAF complex in dcKO_hGFAP-Cre mutants led to decreased numbers of OPCs in prenatal cortex. (A) Coronal brain sections of hGFAP-Cre, Rosa-tdTom transgenic embryo revealed a cortex-specific Cre

recombination pattern at E15.5 (Nguyen et al., 2018). (B) Gene enrichment set analysis (GSEA) showing the set of genes linked to OL development highly depleted in P3 hGFAP-Cre dcKO cortices. (C) Selected astrocyte- enriched genes were upregulated; whereas, neuron-, OL-specific- genes were downregulated in dcKO cortices at P3. (D) Immunostaining of PDGFR (green) in coronal section of control and dcKO_hGFAP-Cre brains at P3 as a representative downregulated OL-related gene in dcKO pallium. (D`) Quantitative analyses comparing the percentage of the PDGFR+ per total cortical cells (DAPI+) in the control and dcKO pallium at P3. Data are presented as means ± SEMs (***p < 0.001), experimental replicates (n) = 4. Abbreviations: dTel, dorsal telencephalon; vTel, ventral telencephalon. Scale bars = 200 µm (A), (D).

3.2. Expression of BAF155 and BAF170 in OL lineage

During the embryonic oligodendrogenesis highly proliferative oligodendrocyte precursor cells (OPCs) undergo either symmetric self-renewing to give rise newly generated PDGFR+OPC pairs or exit the cycle to differentiate into immature oligodendrocytes (iOLs) (Nakatani et al., 2013). Finally, the iOLs eventually develop to mature oligodendrocytes (mOLs) to myelinate neuronal axons. To examine the developmental state of the BAF155/170 + cells in the OL line, we co-stained BAF 155/ 170 with the Olig2 universal marker that is expressed through all the cells of the OL lineage. BAF complex staining was labelled the majority of the Olig2+ population cells (BAF155/BAF170+/Olig2+) (Fig. 7A, B). The percentage of BAF155/170^{high+} among Olig2+ OLs in the presumptive striatum at E15.5 is 85.92% and 88.17%, respectively, proposing that BAF complex is implicated in developing of the oligodendroglial lineage along the embryogenesis (Fig. B, F). The transition between proliferation and differentiation is a key player in regulating the typical development of the oligodendroglial cell line (Matsumoto et al., 2016). Based on the transcriptional OL heterogeneity, we identified the OL population expressing

BAF155/BAF170 using double immunohistochemistry (IHC) with antibodies for BAF155/BAF170 and cell-type-specific markers on brain sections at embryonic day (E15.5) (Fig. 7C-E). The results showed that BAF complex is expressed in the pattern of cells that are included in sequential steps of OL development as demonstrated in Fig. 7C-E, involving glial cells (BAF155/BAF170+/Sox9+, Fig. 7C), in OPCs (BAF155/BAF170+/PDGFR+, Fig. 7D), in iOLs (BAF155/BAF170+/Sox10+, Fig. 7E). The proportion of BAF 155/170^{high+} in Sox9+ glial cells, PDFGR+ OPCs, and Sox10+ iOLs are 96.97% , 83.67% and 85.19%, respectively (Fig. 7F). To sum up, our gene expression pattern analyses corroborated that most of the OL population are highly expressing the BAF complex among the successive OL development.

A
E15.5
WT
BAF155/BAF170- DAPI
B Olig2
C Sox9
D PDGFR
E Sox10
B1
B2
F
BAF comp. High
BAF comp. Low
% Sox10+
% PDGFR+
% Sox9+
% Olig2+
0
20
40
60
80
100

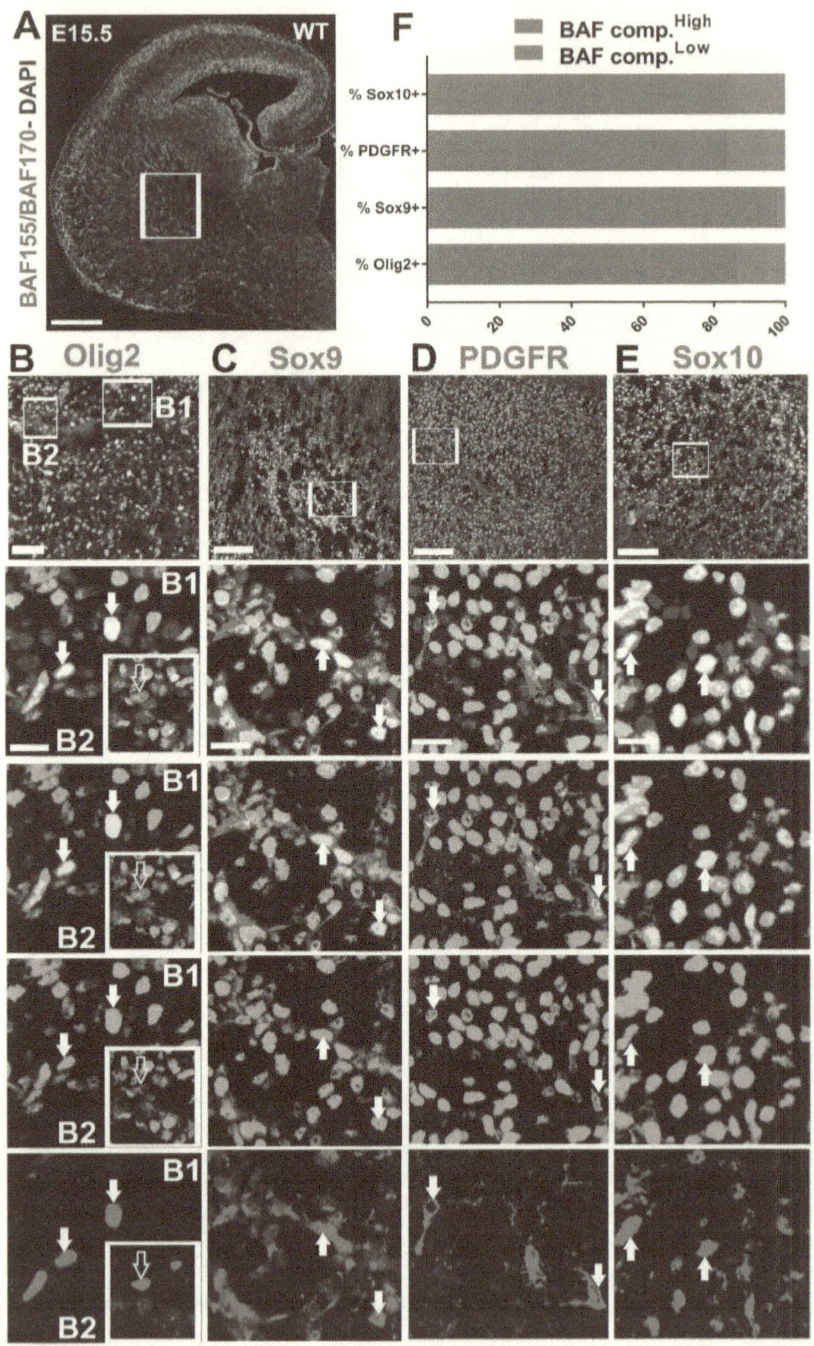

28

Figure 7. Expression of scaffolding BAF complex subunits (BAF155 and BAF170) in the oligodendroglial lineage in the ventral telencephalon.
(A-E) BAF155 and BAF170-expressing cells were characterized in coronal sections of the forebrain at E15.5 by double-label immunofluorescence microscopy using an antibody for BAF155/BAF170 (green) in combination with antibodies against the following marker proteins (red): Olig2 (B, pan OLs), Sox9 (C, glioblasts), PDGFR (D, OPCs), Sox10 (E, iOLs). Filled arrows refer to the oligodendrocytic lineage strongly expressing BAF155/170, whereas, the empty arrows indicate cells exhibiting low expression of the BAF155/170. (F) Estimate of the percentage of the indicated marker protein + cells that are double-positive for BAF155/BAF170 and the marker protein. Experimental replicates (n) = 3. Scale bars = 200 µm (A), 25 µm (B), 50 µm (C-E).

3.3. Conditional deletion of BAF155 and BAF170 in OL lineage results in a global diminution of OL pool in developing forebrain

Based on RNA sequencing data, the elimination of the cortical chromatin remodelling BAF complex mediated by hGFAP-Cre led to an obvious reduction of the OPCs number along the cortex of the adult dcKO_hGFAP-Cre mice. To ascertain the implication of the BAF complex in the embryonic OL production, we conducted the same floxed alleles that previously used, but it was mediated by Olig2-Cre promoter, creating BAF155 cKO, BAF170 cKO, BAF155/BAF170 dcKO mouse line. Unlike the hGFAP-Cre line, Cre activity commences in the developing pallium at E15.5 onward (Nguyen et al., 2018), the Olig2-Cre is active in the embryonic subpallium as early as E12.5. To estimate BAF155/BAF170-dependent regulation of OL production, we initially analysed the universal OL lineage marker (Olig2) (Silbereis et al., 2014) in E15.5 developing brains of BAF155/BAF170 dcKO and control mice. Evaluation of the oligodendroglial pan in the striatum displayed an approximate 50.54% reduction of Olig2+ expressing cells in dcKO mice compared with their control littermates (Fig. 8A, A'). To confirm these observations, we implemented ISH experiments for the pan OL mRNA transcripts (Olig2) at different embryonic phases in developing forebrains of the control and dcKO animals (Fig.8B-D, 9A-C). The ISH analyses exhibited that, unlike controls, the Olig2 mRNA transcripts were downregulated in dcKO subpallium at E15.5 (Fig.8B-D, 8B`-D`). At the advanced developmental stage, the Olig2 expression has dramatically dwindled in the ventral and the dorsal forebrain in the BAF complex-deficient mice (Fig. 9A-C). Additionally, the quantitative analyses indicated that the number of the Olig2+ OLs was strongly diminished along the entire forebrain in the mutants compared to the wild-types (Fig. 9A`-C`). To provide additional support, we further carried out ISH experiments with coronal sections from rostral to caudal telencephalon of to quantify Olig1 transcripts (Olig1). Consistent with Olig2 marker results, the expression of Olig1 showed a global reduction in the E15.5 dcKO brains (Fig. 10 A-C, 10A`-C`). Together, these findings evidence that the BAF155 /BAF170 subunits are linked to govern the oligodendroglial cell fate in the developing telencephalon.

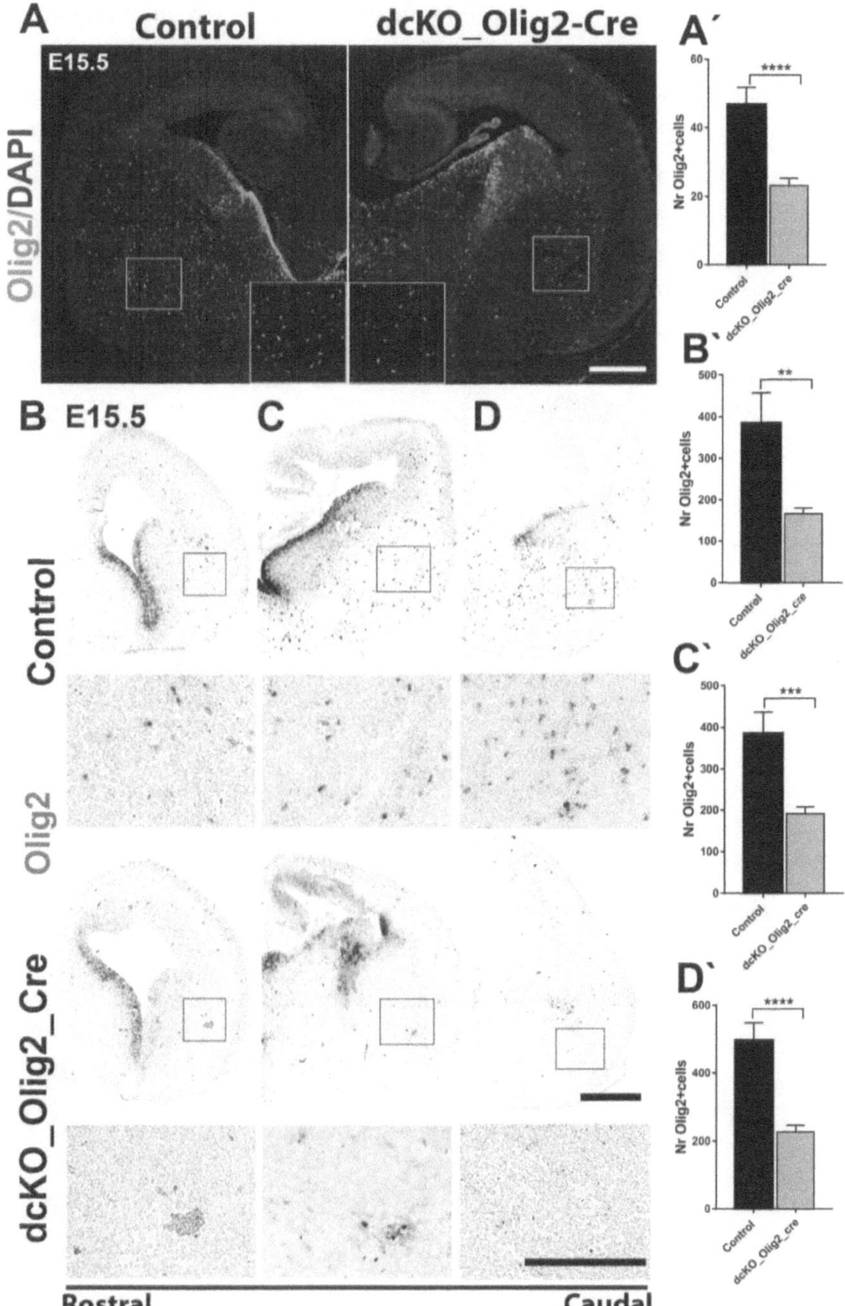
A
Control
dcKO_Olig2-Cre
A´
E15.5
Olig2/DAPI
Nr Olig2+cells
****
Control
dcKO_Olig2_cre
B
E15.5
C
D
Control
B`
Nr Olig2+cells
**
Control
dcKO_Olig2_cre
Olig2
C`
Nr Olig2+cells
***
Control
dcKO_Olig2_cre
dcKO_Olig2_Cre
D`
Nr Olig2+cells
****
Control
dcKO_Olig2_cre
Rostral
Caudal

Figure 8. Reduction of the Olig2+ OL pool in developing forebrain of the dcKO_Olig2_Cre embryo. (A, A`) IHC (A) and quantitative (A`) analyses indicate that the dual loss of BAF155 and BAF170 leads to a diminished number of Olig2+ cells. (B-D) ISH using Digoxigenin- labeled RNA probes that specifically label OL population (Olig2) of control and dcKO_Olig2_cre mouse along the rostral-caudal axis of the telencephalon at E15.5. Magnified images of the delineated zones located in the ventral subpallium show comparison of the number of the Olig2+ expressing cells in controls and mutants. (B`- D`) Statistical analyses indicate significant depletion of Olig2+ cells across the telencephalon rostral-caudal areas in the dcKO_Olig2_cre mutants. Data are expressed as means ± SEMs of replicates (n) = 6 (**p ≤ 0.01, ***p ≤ 0.001, ****p ≤ 0.0001). Scale bars = 200 µm (A), 100 µm (B).

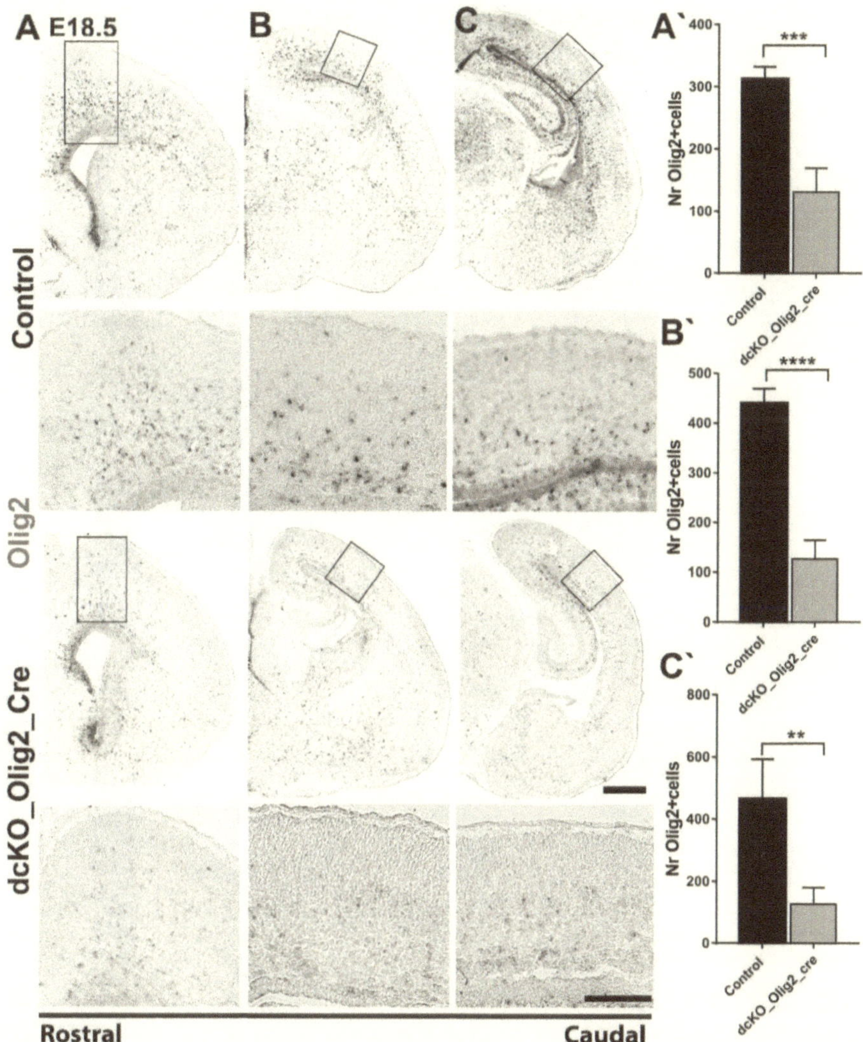

Figure 9. Embryonic murine forebrains affected by BAF complex deficiency reveal depleted numbers of cells expressing a pan-oligodendroglial marker. (A-C) ISH for RNA probes that specifically label OL population (Olig2) of WT and mutants along the forebrain coronal sections at E18.5. Higher magnifications of the corresponding pallium show reduction of the number of the Olig2+ expressing cells in mutants. (A`-C`) Statistical quantifications indicate significant depletion of Olig2+ cells across the telencephalon rostral-caudal areas in the dcKO_Olig2_cre mutants.

Data are shown as means ± SEMs of replicates (n) = 6 (**p ≤ 0.01, ***p ≤ 0.001, ****p ≤ 0.0001). Scale bars = 100 µm.

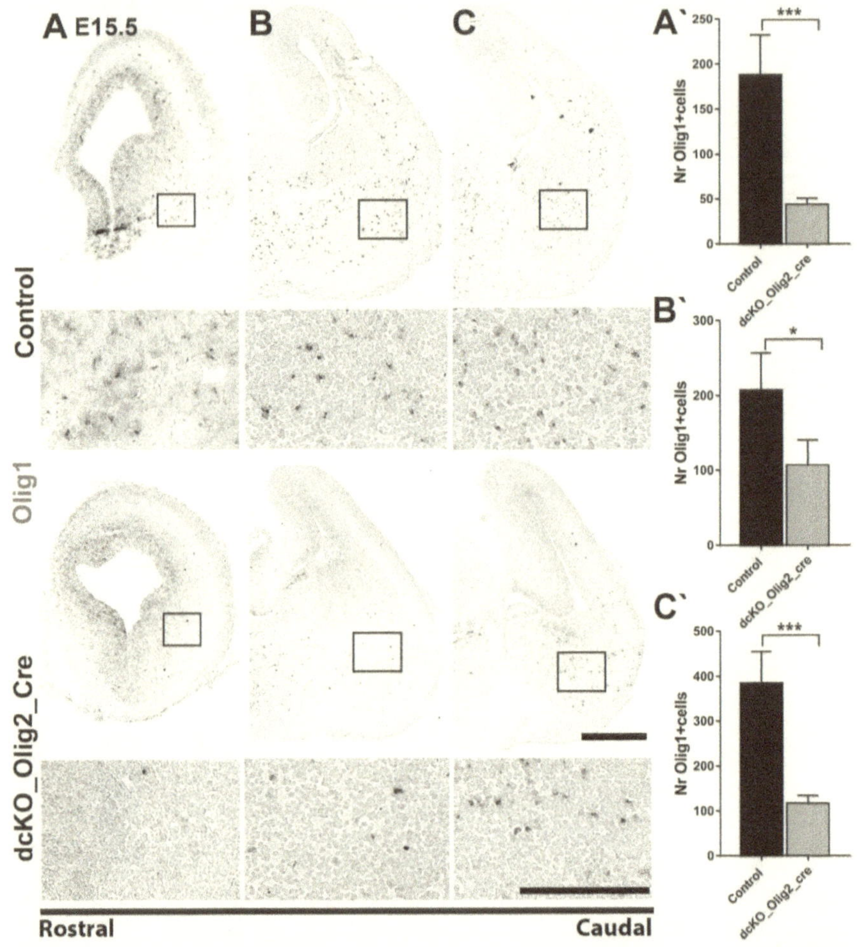

Figure 10. BAF155/170 deficient brains display decreased numbers of Olig1 expressing OLs. (A-C) Expression of Olig1 was detected by ISH on brain sections at E15.5 controls and dcKO_Olig2_cre mutants. Enlarged views of the delineated zones located in the ventral subpallium show comparison of the number of the Olig1+ expressing cells in controls and mutants. (A`-C`) Statistical analyses indicate significant depletion of Olig1+ cells across the telencephalon rostral-caudal areas in the mutants. Data are expressed as means ± SEMs of replicates (n) = 6 (*p ≤ 0.05, ***p ≤ 0.001). Scale bars = 100 µm.

34

3.4. Reduction of proliferation capacity in BAF complex-deficient OPCs

Given the critical role of the cell cycle regulators in the proper oligodendrocytic lineage progression, we sought to investigate the capacity of the OPCs proliferation in the BAF155/BAF170 deficient mice. Under normal conditions, the committed OPCs to divide either symmetrically to produce a pair of cycling OPCs or asymmetrically to give rise to one proliferated OPC and one differentiated oligodendrocyte (Nakatani et al., 2013; Sugiarto et al., 2011; Zhu et al., 2011). In contrast to subpallium born neural cells, OPCs maintain their proliferative capacity beyond the specification of the OPCs (Emery and Lu, 2015; Petryniak et al., 2007). Fig. 11A illustrates the reduction of the number of PDGFR+ OPCs in the BAF155/BAF170 dcKO striatum when compared with the controls. Intriguingly, the analysis of the Ki67 expression, which marks the OPCs actively in the cell cycle, revealed a reduced percentage of PDGFR+/ Ki67+ proliferating OPCs per total PDGFR+ in the dcKO ventral forebrain compared with the wild-types (Fig. 11B, D). To phenotypically confirm that the BAF155/BAF170 loss of function give rise to alteration of the OPCs proliferative fate, we incorporated *in vivo* DNA labelling of the S phase cycling OPCs with the thymidine analogue (IdU) pulses (Fig. 11C). The PDGFR + OPCs had much less prominent IdU signals, signifying the absence of IdU integration into the DNA replication. The ratio of the PDGFR + IdU+ relative to the total OPCs number indicated a significant reduction in the cycling rate of the mutant OPCs relative to wildtypes (Fig. 11C,D) Additionally, we detected no apoptotic defects in the BAF155/BAF170 deficit striatum by scrutinized IHC analysis of cleaved caspase 3 in different coronal brain sections from E13.5 to E18.5 (Fig. 11E, at E15.5 as an example). To sum up, these data prove dysregulation of OPCs proliferative rate caused by the absence of BAF complex.

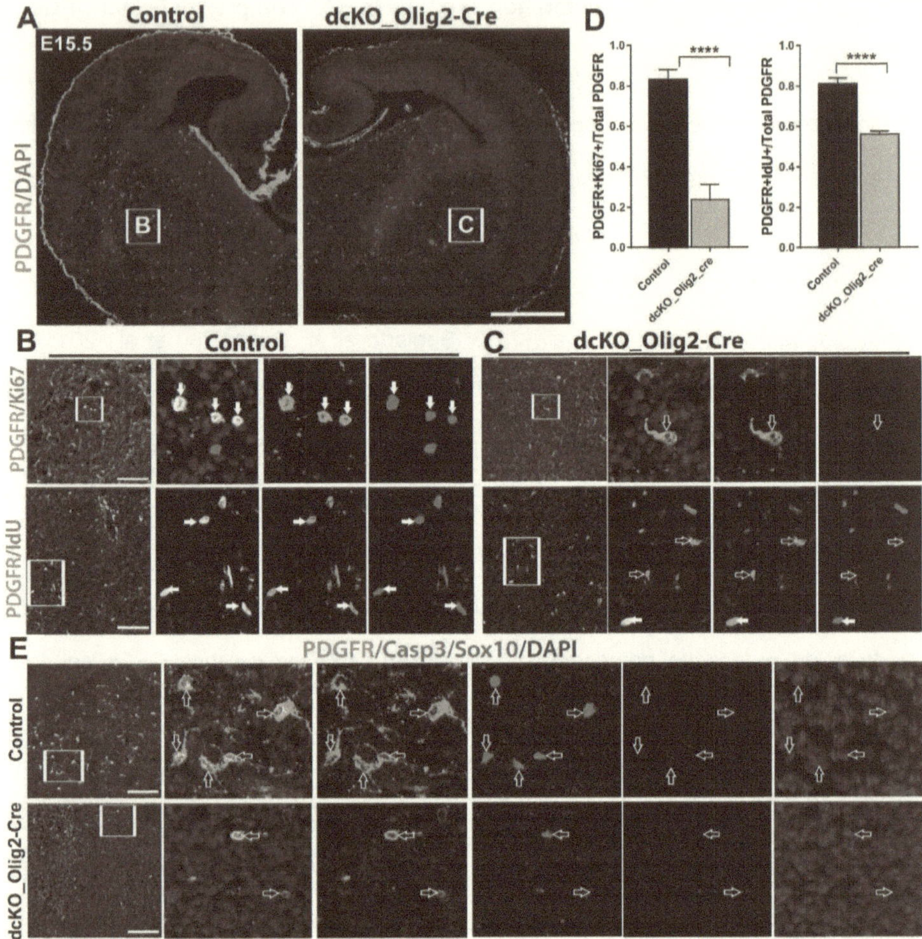

Figure 11. Conditional removal of BAF155/BAF170 leads to an impaired proliferative capacity of OPCs. (A) Immunostaining of PDGFR labels OPCs depicting reduction of the number of PDGFR+ expressing cells in dcKO forebrains compared to the controls. (B, C) Confocal images represent the outlined regions of the ventral telencephalon for double IHC for antibodies against PDGFR/Ki67 (B) and PDGFR/IdU (C) that marks proliferating progenitors. Filled arrows show PDGFR OPCs that are positively expressing Ki67 or IdU, and the empty arrows point to PDGFR OPCs, which are negative with Ki67 or IdU. (D) Statistical analyses revealed significant reduction of the ratio PDGFR+Ki67+/total PDGFR+ as well as the ratio PDGFR +IdU+/total PDGFR+ ; indicating diminution of the proliferative rate of OPCs

in the dcKO_Olig2_cre subpallium compared to controls. (E) Magnified images denote the defined regions of the ventral telencephalon for triple IHC for antibodies against PDGFR (green), Sox10 (violet) that marks differentiated OL and Casp3 (red) that labels cells induced to die of control and mutant E15.5 forebrains. Empty arrows refer to PDGFR+Sox10+ cells, which indicate no expression of Casp3. Data are presented as means ± SEMs (****p ≤ 0.0001), experimental replicates (n) = 6. Scale bars = 200 µm (A), 50 µm (B, C, E).

3.5. Expression of iOLs and mOLs markers are severely diminished in BAF155/BAF170-deficient subpallium

Brg1-dependent BAF complex had previously been reported to be a critical key player in the stimulation of the OPC specification to a regulatory player in the oligodendroglial terminal differentiation, without impact on the OPCs proliferation, survival, and migration (Bischof et al., 2015). Typically, Sox10 expression found in OPCs and its expression persists till the end of the iOLs differentiation process (Finzsch et al., 2008; Kessaris et al., 2006; Liu et al., 2007; Stolt et al., 2002). To find out how loss of the BAF complex function affects the progression of the successive oligodendroglial differentiation step, we accomplished ISH of the Sox10 at E15.5 and E18.5 in embryonic mice forebrains of the dcKO and control animals (Fig. 12A-C, 13A-C). In E15.5 wild-type mice, the number of Sox10 + cells displayed a significant decrease in the ventral forebrain of the BAF155/BAF170 dcKO mice (Fig. 12A-C, A`-C`). At E18.5, Sox10 expression was sternly reduced in the cortical and subcortical telencephalon, demonstrating the disruption of the oligodendroglial differentiation (Fig. 13A-C). The quantitative scrutiny revealed a substantial reduction of the Sox10 expressing cells along the forebrain coronal sections (Fig. 13A`-C`). Then, we tested whether the diminishing in Sox10 expression translates to alteration in the maturation of the OLs. Using ISH for the myelinated OL marker proteolipid protein (PLPDM-20), we assessed PLP transcript along the murine forebrains. In control forebrains, No PLP + cells were detected at the rostral group (olfactory bulb is not included) (Fig. 14), whereas few clusters of strongly expressed PLP + mOLs were appeared among medial and caudal levels/ groups in a confined subcortical regions (Fig. 14). In BAF complex deficit forebrains, lesser number of PLP + mOLs was detected at the corresponding sections, particularly medial groups, relative to their control littermates (Fig. 14). These results indicated that differentiation and maturation are severely affected in absence of BAF complex. In summary, our findings demonstrate BAF155 and BAF170 cooperatively control different steps during oligodendrogenesis, including OPC proliferation, differentiation and maturation.

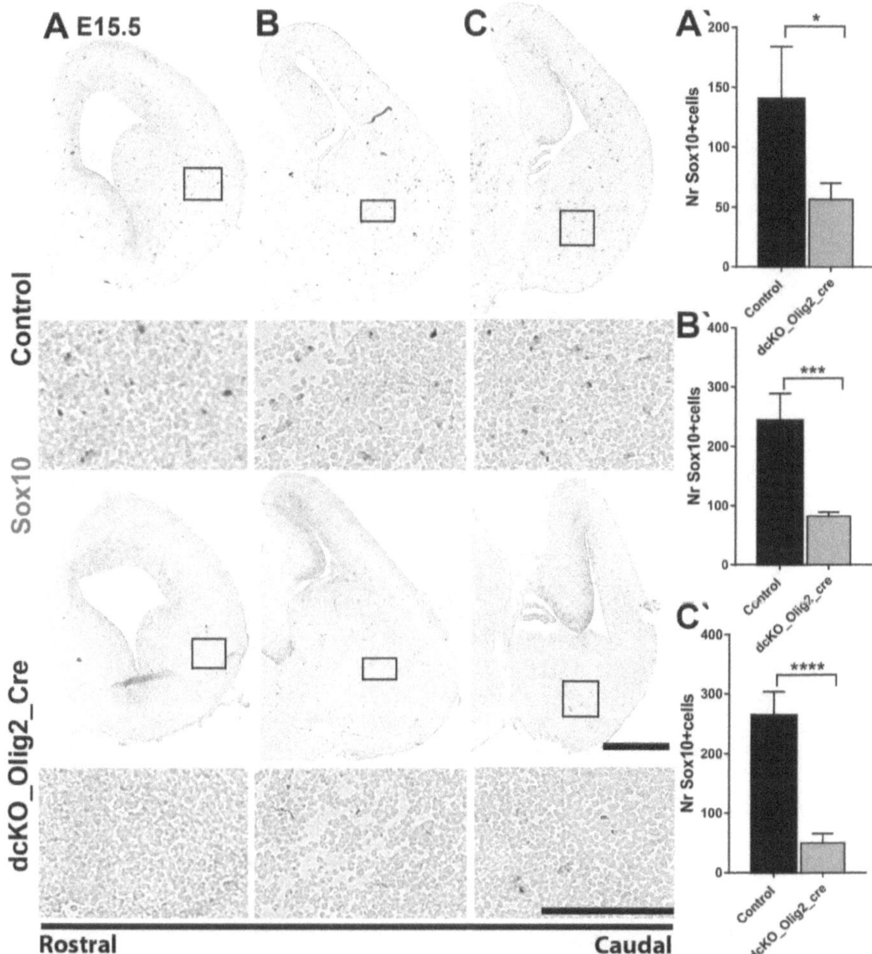

Figure 12. BAF155/170 deficient brains depict diminished numbers of Sox10 expressing iOLs at E15.5. (A-C) Expression of Sox10 was detected by ISH on rostral, medial, and caudal coronal brain sections at E15.5 controls and dcKO_Olig2_cre mutants. Higher magnifications of the delineated zones located in the ventral subpallium show comparison of the number of the Sox10+ expressing cells in controls and mutants. (A`-C`) Arithmetical analyses indicate significant depletion of Sox10+ cells along the rostral-caudal axis in the mutant forebrains. Values are expressed as means ± SEMs of replicates (n) = 6 (*p ≤ 0.05, ***p ≤ 0.001, ****p ≤ 0.0001). Scale bars = 100 μm.

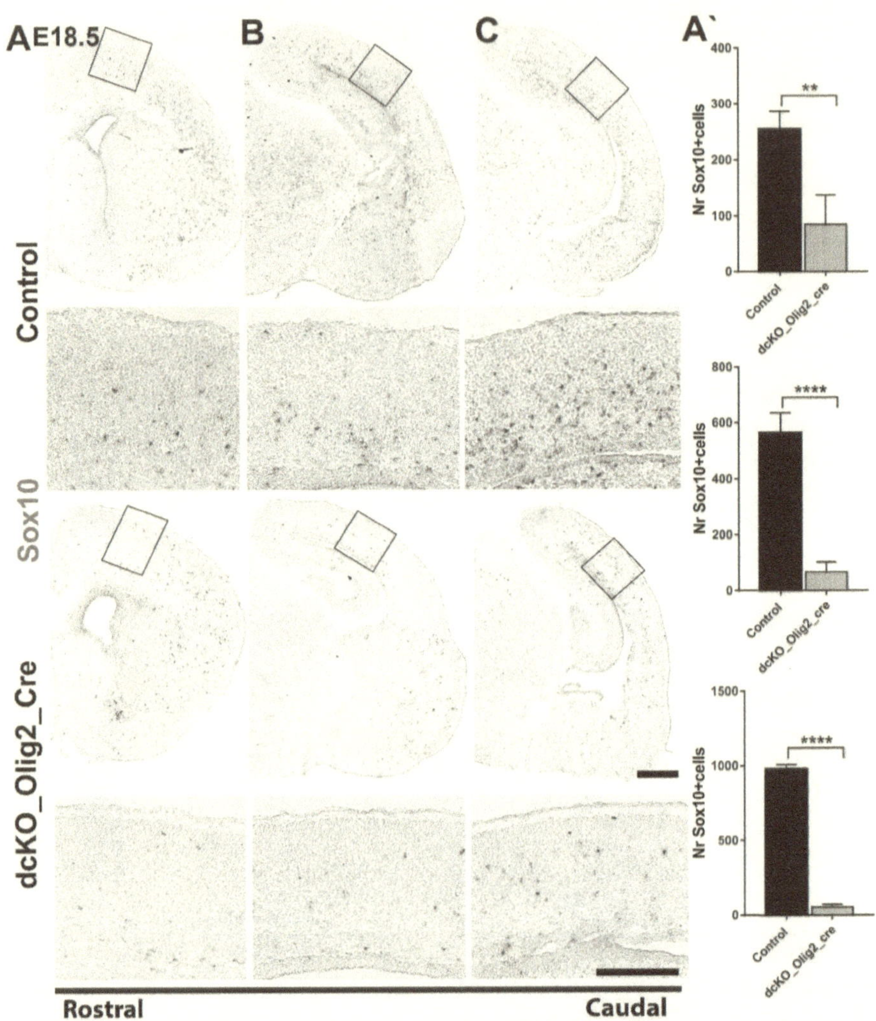

Figure 13. Loss of BAF complex causes significant reduction of iOLs in the developing murine forebrains. (A-C) ISH employing riboprobe that marks immature differentiated OLs Sox10 of wild-type and dcKO_Olig2_cre mouse along the forebrain coronal sections at late embryonic phase E18.5. Higher magnifications of the corresponding pallium display depletion of the number of the Sox10+ expressing cells in controls and mutants. (A`-C`) Arithmetical quantifications indicate significant depletion of Sox10+ cells along the forebrain rostral-caudal regions in the

40

dcKO_Olig2_cre mutants. Values are expressed as means ± SEMs of replicates (n) = 6 (**p ≤ 0.01, ****p ≤ 0.0001). Scale bars = 100 µm.

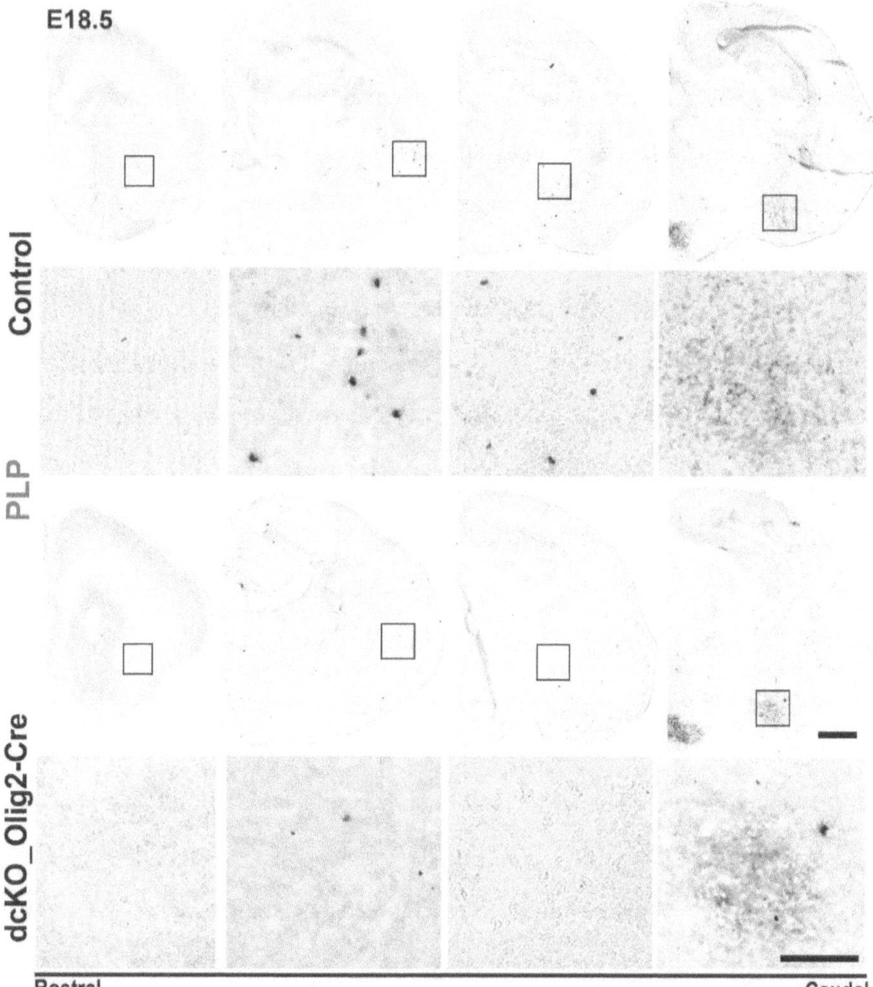

Figure 14. Removal of BAF complex causes abnormal maturation of OLs in the developing murine telencephalon. ISH employing riboprobe (PLP) that marks mOLs of wild-type and dcKO_Olig2_cre mouse along the rostral-caudal axis of telencephalon at E18.5. Magnified views of the delineated zones located in the ventral subpallium highlight the PLP+ expressing cells in controls and mutants. Experimental replicates (n) = 5, Scale bars = 100 µm.

3.6.   Cre fate mapping of cell lineages with Olig2+ history

Fascinatingly, the multipotent subpallial progenitors, which characterized by Dlx2 and Olig2 expression represented key elements in regulating the fates of a multitude of cell lines along with the forebrain development, including the switch between neural and glial lineage fates (Petryniak et al., 2007). To identify the progeny of these dividing cells, which corresponded to the Olig2 expression history, we made use of two knock-in mouse lines with Cre under the mediation of the Olig2 reporter. Thus, Olig2-Cre and BAF155/170dcKO_Olig2-Cre reporters were bred with Rosa26-LSL-Tdtomato reporter mouse model (Ai9), generating Olig2-Cre+/Rosa26-tdTomato and BAF155/170dcKO_Olig2-Cre+/Rosa26-tdTomato mice strains that labelled Olig2 Cre+ progeny with the red fluorescent protein.

We first examined the versatility of the fluorescent tdTomato signal, the whole mount brains of the generated progenies (Olig2-Cre+/Rosa26-tdTomato and BAF155/170dcKO_Olig2-Cre+/Rosa26-tdTomato) were employed for direct fluorescence from tdTomato images (Fig. 15A, B). At E15.5, heterozygote forebrains depicted strong red tdTomato signals in the ventral side, whereas, the fluorescence intensity in the dorsal side showed lesser fluorescent emission than the ventral view of the forebrains (Fig. 15A2, A2). dcKO brains exhibited stronger tdTomato expression in the ventral subpallium, but the lesser expression in the pallium compared to heterozygote brains (Fig. 15B2, B2`). To provide additional support, we employed heat maps for the respective brains analysing native tdTomato signals into a series of visualized colours (Fig. 15A3, A3`and B3, B3`). Consistently, the highest expression of tdtomato appeared in the ventral telencephalon in the E15.5 dcKO brains (Fig. 15B3, B3`). Based on the histological analysis, the characteristic features of the Olig2-Cre+/Rosa26-tdTomato mouse model (control) at E15.5 have been phenotypically described in fig. 15C, tdTomato staining was intensively labelled the GE (VZ and SVZ)/ POA regions and zona limitans intrathalamica (ZLI). A vast number of tdTomato+ cells were populated the mantle stratum and the striatum of the ventral subpallium. Additionally, two migratory streams of tdTomato +cells have been arranged in the neocortex; the first set was observed via the intermediate zone (IZ) and subventricular zone (SVZ), while the second set was organized along the marginal zone (MZ). Unlike controls, the characteristic features of the dcKO_Olig2-Cre+/Rosa26-tdTomato mice demonstrated the abnormal spatial distribution of the tdTomato +cells in the telencephalon. Remarkably, we detected accumulation of the Tdtomato+ cells in the developing regions of the subpallium landscapes; including GE (VZ and SVZ), mantle; however, few numbers of cells immigrated in the cortex. We then investigate the ordinary identity of the migrating GE-derived cell populations that had an ancestor Olig2 profile, using Olig2-Cre+/Rosa26-tdTomato mouse line. Notably, the cortical tdTomato fluorescent cells defined different MGE-derived differentiated INs (84.84% tdTomato+/Sox6+/total tdTomto+), fewer overlapped with Olig2+ immunoreactivity (11.36% tdTomato+/Olig2+/total tdTomto+) and BLBP staining that marks in principle the astrocytic progenitors (9.80 % tdTomato+/BLBP+/total tdTomto+) (Fig. 15D, E). Collectively, the outputs Cre mapping experiments propose that the BAF155/170dcKO_Olig2-Cre mouse model acts as a good candidate to inspect the role of the chromatin remodelling BAF complex in governing the cell fate decision among neural and glial cell lineages.

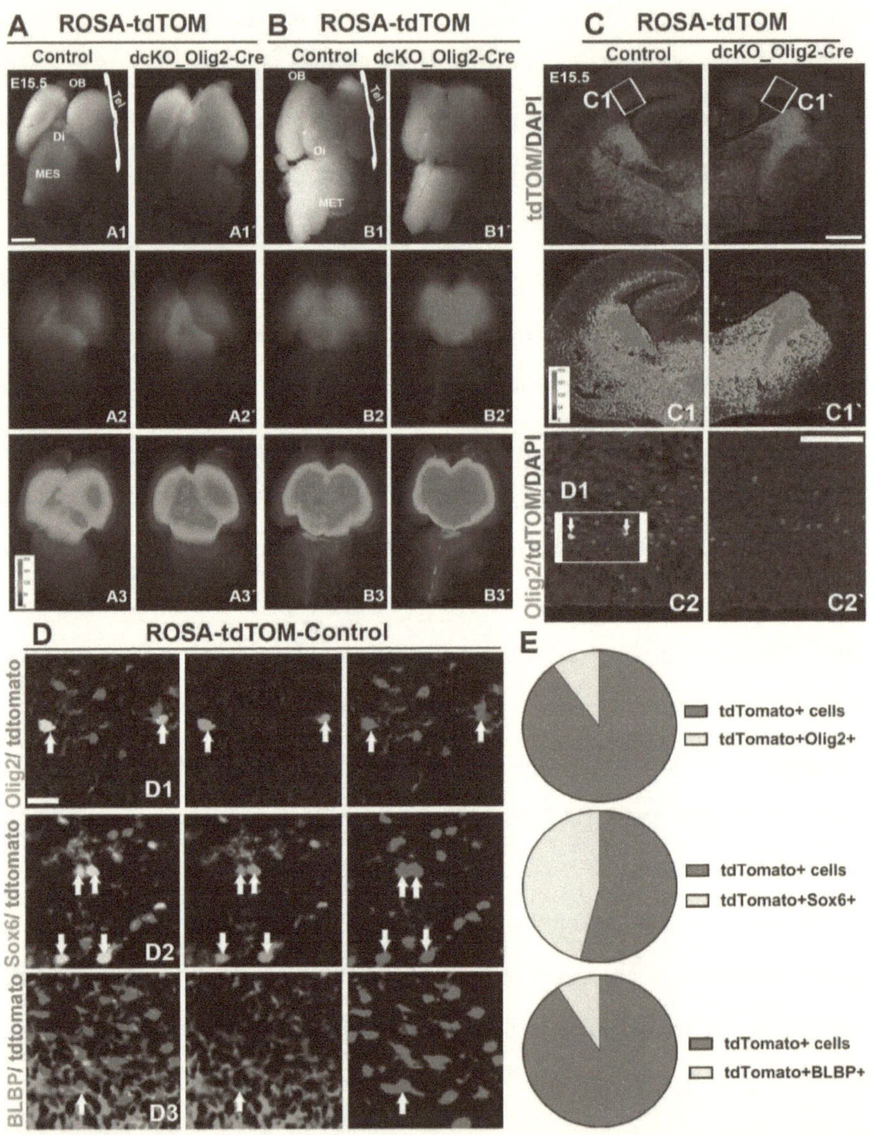

Figure 15. Fate mapping of target cells expressing tdTomato in the murine embryonic Olig2-Cre tdTomato brain. (A) (A1, A1′) Bright field images of whole-mount murine brains (dorsal view) from E15.5 controls and dcKO_Olig2_cre mutants. (A2, A2′) Direct fluorescence of whole-mount mouse brains control and dcKO. (A3, A3′) Heat map of the of the respective brains depicting the gradient of

the native tdTomato intensity, dark violet colour refers to the highest intensity (lies in the ventral side) and blue colour (locates in the dorsal side) indicates the lowest intensity according to the heat map calibration bar. (B) (B1, B1′) Bright field images of whole-mount murine brains (ventral view) from E15.5 controls and dcKO_Olig2_cre mutants. (B2, B2′) Direct fluorescence of whole-mount mouse brains control and dcKO. (B3, B3′) Heat map of the of the corresponding brains identifies the expression of the native tdTomato signals, dark violet colour refers to the highest signal (lies in the ventral side) and blue colour (locates in the dorsal side) indicates the lowest signal. (C) Direct fluorescence localization of the native tdTomato signals on medial coronal sections of E 15.5 control and mutant of murine forebrains. (C1 and C1`) Heat map identifies the native tdTomato intensity of the corresponding brain sections, dark violet colour refers to the highest signal (lies in the ventral side) and blue colour (locates in the dorsal side) indicates the lowest intensity. (C2 and C2`) magnified views were taken from the outlined areas in the developing Ctx, designates the diminution of the number of tdTomato expressing cells in mutant cortex compared to mutant. (D) (D1 - D3) Magnified images identifies tdTomato signals (red channel) which co-localized different cell type specific markers (green channel), involving (D1) Olig2 marks OLs, (D2) Sox6 labels IN cell lineage (MGE born INs) and (D3) BLBP defines astrocytic population in the medial coronal sections of forebrain at E15.5 wild type. (E) Quantification analyses for cells expressing specific marker (Olig2 or Sox6 or BLBP) and are positively expressed tdTomato in striatum at E.15.5. Scale bars = 300 μm (A, B), 200 μm (C1`), 50 μm (C2`), 10 μm (D). Abbreviations: OB, olfactory bulb; Tel, telencephalon; Di, diencephalon; MES, mesencephalon; MET, metencephalon.

3.7.   Expression pattern of BAF155 and BAF170 in IN lineage

Based on the expression of molecular markers, the subpallium is divided further into: LGE, MGE, CGE, Sep and POA (Hernández-Miranda et al., 2010). In rodents, INs are presented in the postnatal mouse pallia, however, majority of them are generated from the developing MGE (Chen et al., 2017). Beyond the neural fate acquisition, MGE -derived new-born INs are impelled with a succession of differential and migration programs, which drive them to migrate tangentially to the pallium (Poitras et al., 2007; Tamamaki et al., 1999; Wichterle et al., 1999; Xu et al., 2008). Indeed, the neuroblasts are subjected to combinatorial key factors that are crucial to neuronal differentiation, patterning and migration. Here, we characterized the expression of the BAF155/170 during the subpallial neurogenesis; double IHC was labouring different neural developmental cell markers, involving Nkx2.1 and Sox6 with the BAF155 and BAF170 antibodies (Fig. 16A,B). Fate mapping studies reported that MGE - derived Nkx2.1 progenitors produce diverse cell types; involving inhibitory INs, OLs and astrocytes, which populate the pallial and pallidal telencephalon (Corbin et al., 2001; Kessaris et al., 2006; Marín and Rubenstein, 2001; Minocha et al., 2017). Figure (16A,A`) depicted strong labelling of the two BAF subunits among the majority of the Nkx2.1+ cells in MGE and globus pallidus (GP) (BAF155/BAF170+/Nkx2.1+). Further immunostaining of the Sox6 that marks migrating INs revealed all Sox6+ INs are also immunoreactive with BAF155/BAF170 (BAF155/BAF170+/Sox6+) at E15.5 and E17.5, suggesting that BAF complex might has an important role in the migration of INs from the subpallial toward the pallium (Fig. 16B, B′). Overall, the gene expression analysis revealed that neural cell populations, which are generated from subpallium, express BAF155/BAF170.

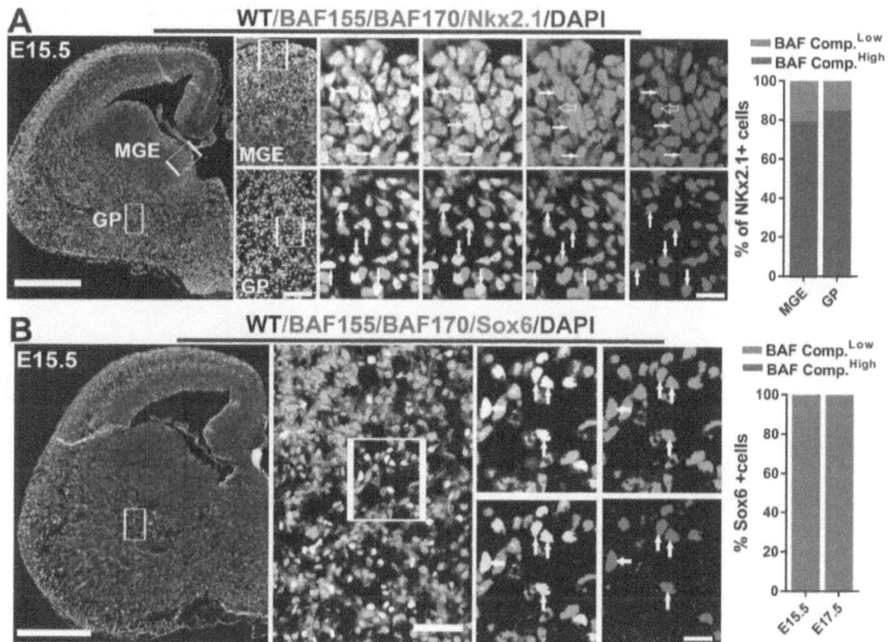

Figure 16. Expression of BAF complex subunits in neural lineage in the ventral telencephalon. (A) Double immunostaining with coronal sections of forebrain at E15.5 and antibodies against BAF155 and BAF170 subunits (in green) and Nkx2.1 (in red) as molecular marker for neural progenitors and post-mitotic neurons wild type. Confocal magnifications of the delineated zones involve MGE and GP. Filled arrows indicate to the IN population markers strongly expressing BAF155/170, whereas, the empty arrows demonstrate cells exhibiting low expression of the BAF155/170. (A′) Ratios of positively expressed BAF subunits per total Nkx2.1positive in MGE / GP at E15.5. Scale bars = 200 µm, 50 µm, 10 µm respectively. (B) Double immunohistochemical staining of BAF 155 and BAF170 subunits (in green) and Sox6 (in red) labels the migrating differentiated MGE derived IN in the medial coronal sections of forebrain at E15.5 wild type. Magnified views of the delineated ventral forebrains illustrate complete strong expression of BAF complex in Sox6 –expressing migrant neurons. (B′) Ratios of positively expressed BAF subunits per total Sox6 positive in striatum at E.15.5 and E17.5. Scale bars = 200 µm, 25 µm, 10 µm respectively. Experimental replicates (n) = 3.

3.8. BAF complex ablation in the MGE progenitors leads to severe reduction of the GABAergic INs in the embryonic pallium

To study the potential function of the BAF complex in the neurogenesis of the subpallial NSCs, we employed correspondingly Olig2 cre reporter for the conditional removal of the BAF 155 and BAF 170 alleles from the NSCs in the ventral telencephalon. As previously mentioned, the activity of the Olig2-Cre commences around E12.5 specifically at the developing subpallium. Of greater importance, GABA is the chief inhibitory neurotransmitter, which mainly organizes the local cortical INs circuits, among their excitatory counterparts (Sultan and Shi, 2018). To assess the effect of the BAF complex in the development of the MGE derived INs, we performed IHC experiments with antibody to label the pan IN marker (GABA) of the control and dcKO developing brains (Fig. 17A). The analysis revealed obvious depletion of the GABA+ cells in the dcKO pallia compared to their control littermates (Fig. 17A,A`). Given the significance of the transcriptional regulator Nkx2.1 in the generation of the IN subgroups in striatum and cortex, we assessed the Nkx2.1 expression in the ventral and dorsal forebrains. ISH was accomplished for Nkx2.1 mRNA transcript at E15.5 and E18.5 (Fig. 17B). At E15.5, ISH staining revealed downregulation of Nkx2.1 expression in the mutant GPs compared to their control littermates. At late embryonic age (E18.5), the expression of Nkx2.1 is normally persisted among the post-mitotic INs in different subpallial regions; such as striatum, and GP (Fig. 17B). Nonetheless, the mutant pallidum showed reduction of number of Nkx2.1+ cells in GP, as well as, abnormal aggregation of the Nkx2.1+ cells in the mutant MGE/POA region. These data propose the inability of the Nkx2.1 cells to emigrate outside the mutant MGEs. To confirm these findings, we carried out double immunostaining for antibodies against Nkx2.1 and Sox6 (Fig. 17C). Of note, the number of Nkx2.1+/Sox6+ neurons were significantly decreased in the mutant GPs relative to their peers (91.73%), suggesting the influence of the BAF complex in controlling the maintenance of the various subclasses of the post-mitotic neurons in the GP (Fig. 17C, C`). In addition, figure 18A depicts low number of Sox6+ INs in the pallium and subpallium of the dcKO mouse brains compared to their littermates at E18.5. Furthermore, the quantitative statistical analyses showed significant reduction of Sox6+ INs in the striatum and severe depletion of Sox6+ INs in the dcKO dorsal telencephalon relative to the controls at E18.5; proposing a disruption in the typical differentiation progress of the neural progenitors in dcKO_Olig2-Cre mice (Fig. 18A, B). Notably, the ratio of the Sox6+ INs in the dcKO striatum versus the dcKO pallium revealed a decrement by 62.77%, suggesting the critical involvement of the BAF complex in migration of the Sox6+ INs toward the dorsal telencephalon (Fig. 18A, B). Overall, these outputs display that BAF155/170-deficient MGE showed aberration of cortical INs development and dysregulation of the migration route of the MGE progenitors.

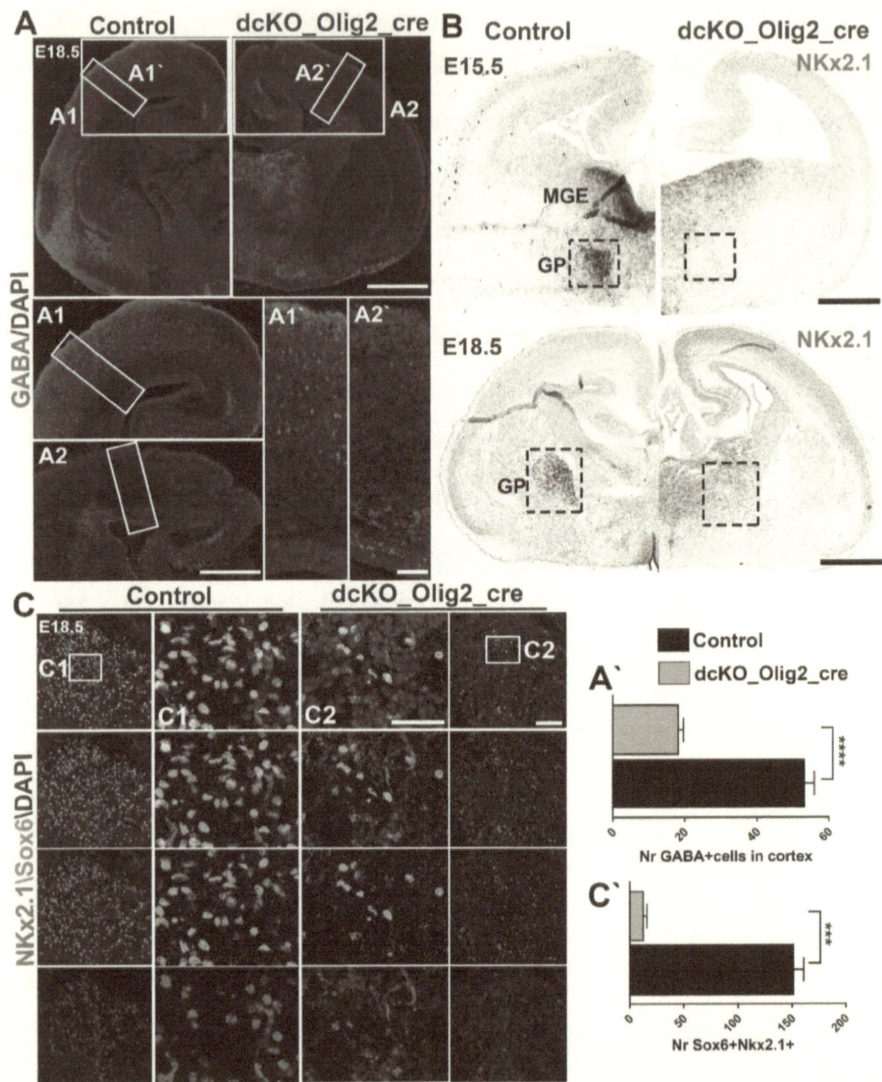

Figure 17. BAF155/170 deficient embryonic forebrain reveals developmental alteration of MGE derived INs. (A) Reduction of GABAergic cortical INs in the dcKO cortices were perceived by means of IHC using GABA antibody as a pan-IN marker in medial coronal brain sections of wild-type and dcKO_Olig2_cre mouse at late developing stage E18.5. (A1, A2) are magnifications of the corresponding cortices in controls and mutants respectively. (B) Expression of Nkx2.1 was detected

employing ISH on medial coronal brain sections at E15.5 and E18.5 controls and dcKO_Olig2_cre mutants. Higher magnifications of the outlined zones located in the GP demonstrate lessening Nkx2.1 expression in controls rather than mutants. (C) Confocal images illustrate the co-localization of the Nkx2.1+ Sox6+ INs by means of double IHC for antibodies against Nkx2.1 and Sox6. (C1, C2) zoom images to visualize the overlapping of Nkx2.1 and Sox6. (A′) Statistical analyses revealed significant reduction of the GABA+ neurons in the dcKO pallia compared to controls. (C′) Arithmetical quantifications of co-expression of NKx2.1 and Sox6 of medial regions in the dcKO mutants compared to controls. Data are shown as means ± SEMs (***p ≤ 0.001, ****p ≤ 0.0001), experimental replicates (n) = 3. Scale bars = 200 μm (A, B),100 μm (A2), 50 μm (A2`, C), 10 μm (C2).

3.9.   Conditional lack of BAF complex affects the neuronal differentiation and migration

Previous studies illustrated the spatial distribution of the Lhx6-expressing cells in the ventral pallidal and the tangentially migrating INs (Alifragis et al., 2004; Lavdas et al., 1999; Liodis et al., 2007). Thus, we used the spatial patterning of Lhx6 expression as a metrical tool for analysing the phenotype of the migratory route of the cortical INs in dcKO mutants. Close inspections of Lhx6 riboprobe expression have been performed among brain sections of two embryonic ages (E15.5, E18.5) (Fig. 18C). At E15.5, we observed a bulk of Lhx6+ expressing INs were accumulated in the mutant MGE/POA regions, whereas few Lhx6+ differentiated INs were detected in the striatal and pallidal areas of mutants relative to the controls (Fig. 18C). As predicted, we have recognised a consequent disruption of the Lhx6+ migrated INs toward the mutant pallia compared to the normal pallia, which revealed two migratory streams of Lhx6+ INs (Fig. 18C, C1,C2). As depicted in figure (18C), E18.5 brain sections exhibited strong decreases in number of the differentiated INs in the dcKO striatum and pallidum associated with clear aggregations of Lhx6+ INs in the MGE/POA areas. Interestingly, a little set of forerunners were perceived to migrate from the ventral telencephalon to the dorsal pallium (Fig. 18C,C3,C4). To sum up, these data indicated that the deletion of the BAF complex in the subpallial NSCs dysregulate the allocation of migratory routes of the cortical INs.

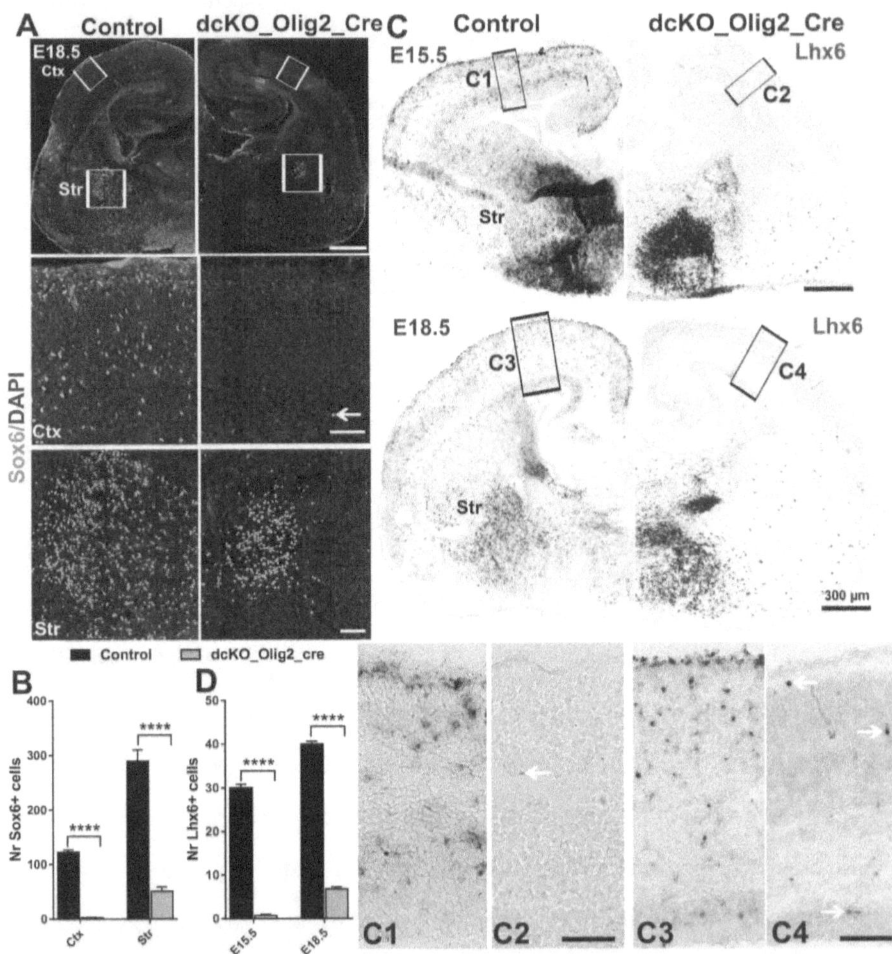

Figure 18. Conditional loss of BAF complex impedes the differentiation and migration of the cortical MGE derived IN. (A) Immunofluorescence staining of wild-type and dcKO_Olig2_cre E18.5 mouse brains to examine expression of Sox6. Confocal images of the corresponding pallial and subpallial regions validate

decrement of the number of Sox6 expressing migrant INs in mutants compared to controls.(B) Calculated quantifications of Sox6+ cells in cortical and striatal regions in the dcKO_Olig2_cre mutants compared to controls. (C) ISH employing riboprobe to label migrating INs (Lhx6) of wild-type and dcKO_Olig2_cre mouse medial coronal of the forebrain sections at E15.5 and E18.5. (C1-C4) Higher magnifications of the outlined zones located in dorsal telencephalon reveal severe reduction of the number of the Lhx6+ expressing cells in mutant cortices compared to control cortices. White arrows refer to few Lhx6+cells in dcKO cortex at E15.5 (C1, C2) and E18.5 (C3, C4). (D) Calculated quantifications of Lhx6+ cells in the dcKO pallia compared to controls at E15.5 and E18.5. Values are shown as means ± SEMs (****p ≤ 0.0001), experimental replicates (n) = 3.Abbreviations: Ctx, cortex; Str, striatum. Scale bars = 200 μm (A), 50 μm (Ctx, Str), 300 μm (C), 50 μm (C2), 100 μm (C4).

3.10. Potential role of BAF complex in NSC fate choice

Dlx2 and Olig2 transcription regulators contest to command the progenitor fate choice toward either neural or glial cell lineages, respectively (Petryniak et al., 2007; Silbereis et al., 2014). To inspect the selectivity of the MGE progenitor's domain, we carried out ISH of the Dlx2 riboprobe transcript on brain sections of the control and dcKO animals (Fig. 19A). Remarkably, ectopic Dlx2 expression exhibited expanded subcortical dcKO MGEs relative to the normal MGEs. Upregulated expression of Dlx2 in dcKO pallidal suggested that the majority of MGE progenitors sustain their cell fate choice toward neurogenesis program. Additionally, we followed the expression of Dlx1 riboprobe, which principally act synergistically with Dlx2 in promoting the neurogenic programs (Fig. 19B). As expected, the subpallial MGE/ POA region was densely populated with Dlx1 signal in the in mutants compared to controls. Overall, we assume that BAF complex–deficit subcortical progenitors may drive their cell fate toward neurogenic program through enhancing the master activators Dlx1/2.

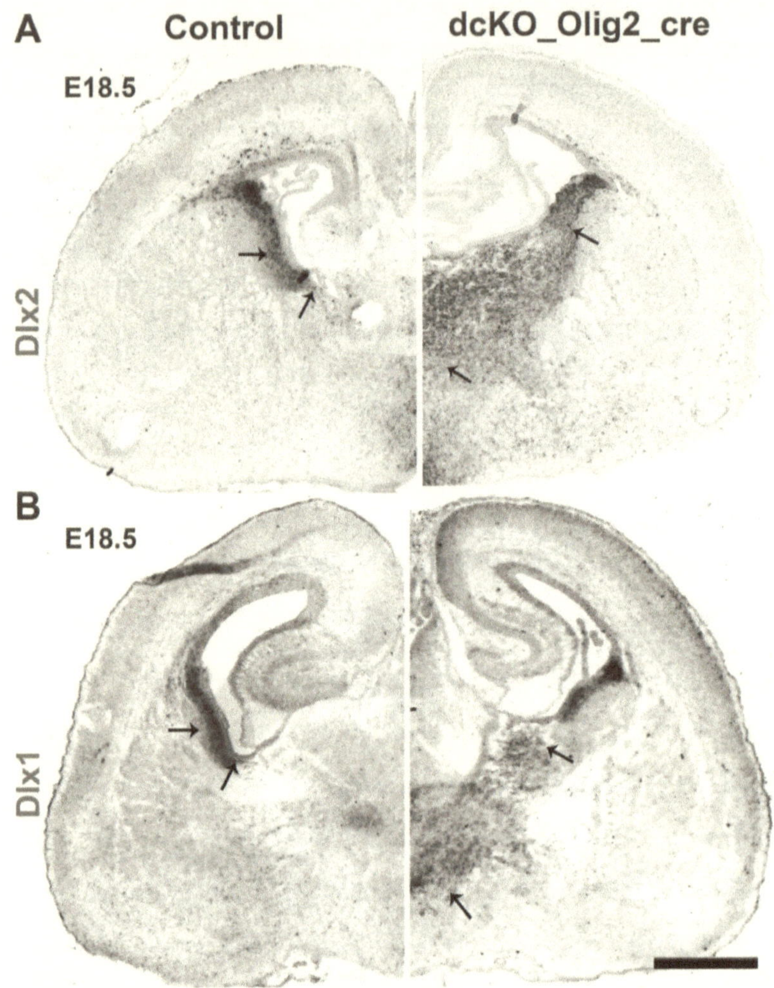

Figure19. Absence of BAF complex might cause inappropriate NSC fate decision manifested by the upregulation of Dlx1/2+ germinative cells in the mutant MGEs. (A) ISH using riboprobe tag the master activators (Dlx2) (A) and Dlx1 (B) of wild-type and dcKO_Olig2_cre mouse medial coronal of the forebrain sections at E18.5. Arrows refer to the expansion of the mutant MGE/POA area rather than wildtypes. Scale bars= 200 μm (A, B).

4. Discussion

Emerged evidences illustrated the implication of epigenetic and chromatin regulation mechanisms in control oligodendroglia development (Bischof et al., 2015; Emery and Lu, 2015; Yu et al., 2013). Further, recent studies have assessed special interests on the neurodevelopmental and psychiatric disorders that are provoked by the abnormal development cortical GABAergic caused by heterozygous elimination of Arid1b – dependent BAF complex in the developing murine brains (Celen et al., 2017; Jung et al., 2017). Herein, we shed light on the significance of the chromatin remodelling BAF complex in regulating the development of the subcortical born glial and neuronal cell lines.

4.1. Expression of BAF155 and BAF170 in the OL lineage

Among the chromatin remodelling complexes, the biological role of the mammalian BAF complex in governing the brain development, including OL production has extensively been investigated (Narayanan et al., 2015; Nguyen et al., 2018; Tuoc et al., 2013; Yu et al., 2013). One example is the catalytic ATPase core subunit Brg1, which is a crucial player in the differentiation of the OPCs, and expressed strongly at the commencement of the oligodendroglia differentiation (Yu et al., 2013). Other BAF complex subunits have been also demonstrated to contribute to enhancing the differentiation program of the OPCs; appertaining, BAF250b, BAF60a, BAF45b, BAF45d (Wu et al., 2009; Yu et al., 2013), and the core subunit Brg1 (Gregath and Lu, 2018; Yu et al., 2013). This latter candidate is highly co-expressed to Brm subunits throughout the OL development (Bischof et al., 2015). In addition to the former manifestations that demonstrated the ability of Brm to substitute Brg1 within BAF components (Hargreaves and Crabtree, 2011; Ho and Crabtree, 2010). Consequently, ablation of Brg1 in OPCs resulted in a mild decrement of the differentiating iOLs by one-third, raising the possibility that the deficiency of Brg1 might partially be recompensed by Brm in the BAF complex of the OLs (Bischof et al., 2015). In the current study, we determined the functional role of the scaffolding subunits BAF155 and BAF170 over the OL lineage progression in the developing telencephalon. We previously proved that BAF155/170 subunits are essential to physically maintain the integrity of BAF complex (Narayanan et al., 2015). Our expression analysis of the BAF155/BAF170 subunits in OL population in the developing forebrain demonstrated that these subunits are presented in entire OL lineage from NSCs to OPCs toward iOLs. This finding implicated that BAF complex plays important roles in multiple steps during OL development.

4.2. BAF155 and BAF170 chromatin remodelling factor subunits are indispensable for OL development

Well documented studies demonstrated the significance of the epigenetic remodelling in controlling the differentiation of OPCs and myelination in the embryonic mammalian brain (Bischof et al., 2015; Copray et al., 2009; Emery and Lu, 2015; Matsumoto et al., 2016; Yu et al., 2013). Moreover, several molecular regulators, which contribute to operating the transcriptional program of iOLs differentiation and maturation are identified; for instance, stage-specific TFs, comprising Olig2, Olig1, Sox10, Zfp191, YY1, and MRF, in addition to minor sets of non-coding RNAs as miR-219 (Emery, 2010; Galloway and Moore, 2016; Li et al., 2009; Lu and Barca, 2012). Of special interest, prior studies focused on the relation between chromatin remodelling activities and Olig2, which act as a pre-patterning key governor in the OL lineage (Bischof et al., 2015; Copray et al., 2009; Emery and Lu, 2015; Matsumoto et al., 2016; Yu et al., 2013). Furthermore, Yu et al. (2013) indicated the prominence of the transcription regulator Olig2 in recruiting Brg1–

based BAF complex to loci for a set of genes at the onset of the OPC differentiation process. Nevertheless, the Brg1-based BAF complex appears to be dispensable for the proliferation of OPCs (Yu et al., 2013). Based on our earlier work, we reported that the entire chromatin remodelling BAF complex is completely abolished by the conditional deletion of BAF155/170 subunits (Narayanan et al., 2015). In the present research, we characterized the phenotypes changes of the conditional knockout of BAF complex driven by hGFAP-Cre in the pallium and by Olig2-Cre in subpallium. At P3, the transcriptional profiling of the BAF complex deficit cortex revealed a downregulated expression of OL genes. *In vivo* analyses confirmed transcriptomic analysis depicting fewer PDGFR+ expressing OLs in dcKO cortices compared to controls. Indeed, these screening data propose that BAF complex play an essential role in operating the oligodendroglial development. Specifically, we elucidated that ablation of BAF complex in embryonic subpallial progenitors led to significant depletion of the proliferative Olig2+ OLs population in the developing mutant pallium and subpallium. Furthermore, the phenotype analyses of the oligodendroglial differentiation and maturation shown that dcKO brains have a severe reduction in the number of Sox10+ iOLs and a subsequent diminution of myelin-forming PLP+mOLs. Interestingly, myelination patterns in the mammalian developing forebrains proceed anatomically from the caudal to the rostral axis (Inder and Huppi, 2000; Jakovcevski et al., 2009; Jakovcevski and Zecevic, 2005; van Tilborg et al., 2018). Of particular focus, few studies described the spatial and temporal expression pattern of PLP in the embryonic murine forebrain, depicting its first appearance in a small number of cells at E16.5 in the hypothalamus, adjacent to the third ventricle at optic chiasm medial level, and infrequent cells in the internal capsule (Hardy and Friedrich, 1996; Spassky et al., 1998). PLP expression area assignments were certified referencing

ISH data from the Allen Institute for Brain Science (Allen Institute's Developing Mouse Brain Atlas (Allen DMBA) and are accessible from (http://developingmouse.brain-map.org). Consistently, our phenotype data demonstrated that at E18.5 more PLP+ mOLs are found caudally in the hypothalamus and internal capsule, however, these clusters of cells are still limited to the level of the optic chiasm of the ventral telencephalon. Our findings revealed depletion of myelin-forming PLP+ mOLs in olig2-dcKO forebrains relatively to controls. Altogether, our findings demonstrate the significance of the BAF 155/170 subunits in supporting the commitment OPCs for differentiation and eventually their maturation. Accordingly, it is conceivable that, at the onset of OL specification, the BAF complex may recruit cascade of genes to enhance the properness OL development pathways. However, we foresee that this assessment of BAF155/170 targets will get attractive attention to upcoming examinations; since these candidates might have beneficial imparts in myelin healing therapy of some demyelinating diseases.

4.3. BAF complex controls the proliferative rate of the OPCs

The impaired development of OL lineage is one of the foremost conditions, causing diverse neurodegenerative disorders (Maki et al., 2013; Ohtomo et al., 2018). Under the pathological circumstances, for instance, multiple sclerosis, numerous damaging cascades are foreseen to be promoted leading to the programmed death of OLs (Maki et al., 2013; Ohtomo et al., 2018). Foremost, we did not detect a significant number of Casp3+ apoptotic cells in BAF155/170 dcKO compared to the controls. Our remarks posited that the OPCs appeared to be relatively quiescent, revealed by no expression of the nuclear proliferation Ki-67 antibody in the majority of the OPCs (PDGFR+Ki67-OPCs) in the BAF complex-deficit mice. Consequently, loss of BAF complex leads to hamper the self-renewal and proliferative activity of OPCs. Nonetheless, a complete understanding of the potential cell fate specification of the NSCs into OPCs will require additional analysis. Altogether, our findings substantiate that the BAF complex plays a pivotal function in the ordinance the action of the transcriptional activators to control the self-renewal and the cell cycle exit of OPCs in the embryonic forebrain.

4.4. Olig2 Cre mapping in glial versus neuronal derived progenitors

Consistent with the conception stated that upon E12.5 (or earlier) the developing Olig2+ progenitors, within the telencephalon, specified into neurons, whereas those in advanced stages leaned more to give rise glial cells (Masahira et al., 2006; Ono et al., 2008). In the present study, short-term Cre mapping analysis at E15.5 verified that the majority of migrated Olig2 Cre+ cells, which are labelled by tdtomato in MGE of the embryonic telencephalon identified as INs (Sox6+tdtomato+) in the cortex. Few numbers of Olig2+ expressing OLs were targeted by tdTomato followed by the astrocytic lineage which was represented by BLBP expression. With reference to the conception stated that developing Olig2+ progenitors, which labelled by Cre-driven recombination in MGE of the embryonic telencephalon engendered mainly NG2 glia or mOLs, but scarcely astrocytes and neurons. Nonetheless, this is marked contradictory to the multiple cell lineages development from the ventral subpallial pool employing different kind of mouse lines (Boshans et al., 2019; Dimou et al., 2008; Furusho et al., 2006; Masahira et al., 2006; Miyoshi et al., 2007; Ono et al., 2008). Our findings revealed that many ventral subpallial Olig2 expressing progenitor cells differentiated to GABAergic INs in the forebrain. GABAergic differentiation of Olig2+ cells was estimated due to their limited localization in the GE (Butt et al., 2005; Ono et al., 2008; Wichterle et al., 2001). Short-term lineage analysis mediated by Olig2_cre_tdTomato clearly demonstrated the differentiation of Olig2+ cells to the multitude of cell lines in the developing cortex. Additionally, these findings highlight the significance of employing BAF155/170dcKO_Olig2-Cre mouse model in the neural/glial development, as well as cell fate decision among neural and glial cell lineages.

4.5. Expression pattern of BAF155/170 subunits among IN lineage

Gene expression is broadly governed by the transcriptional regulators that are coordinated by the chromatin milieu (Boshans et al., 2019). Recent surveys explored the significance of Arid1b, encoding for BAF250B subunit of the chromatin remodelling BAF complex in brain development and behaviour (Celen et al., 2017; Jung et al., 2017). Interestingly, Arid1b+/− and Arid1b-cKO phenotypes displayed ordinary density and the arrangement of the telencephalon population cell lines, encompassing OLs, astrocytes and pyramidal neurons, excepting the GABAergic IN lineage (Celen et al., 2017; Jung et al., 2017). This latter showed a significant decrement in neuronal density, hence, imbalance of the cortical excitatory and inhibitory circuits (Jung et al., 2017). Since the MGE share an early developmental birthplace of cortical GABAergic neurons with the OLs, we sought to examine whether the scaffolding subunits BAF155 and BAF170 involve in neuron-glial fate choice, in addition to the development of INs within the embryonic forebrain. Our expression pattern data of the BAF155 and BAF170 proteins in MGE at E15.5 revealed predominately expression of these subunits in Nkx2.1 expressing population in MGE (VZ and SVZ). Further, BAF155/170 proteins are co-expressed in the post-mitotic group of Nkx2.1 expressing cells in the GP. Importantly, the transcription regulator Sox6 that controls over the maturation and migration of MGE-derived INs (Batista-Brito et al., 2009) showed *in vivo* expression of BAF155/170.

These outputs emphasize the requirement of the BAF complex for development and migration of the MGE derived INs.

4.6. BAF complex controls the differentiation and migration of the MGE precursor cells

Despite current advances understanding the epigenetic regulation in neurodevelopment, the underlying mechanisms for cortical IN development, in particular, remain poorly understood. The Arid1b haploinsufficiency mouse model showed hallmark of various neurodevelopmental and psychiatric disorders; for instance, intellectual disability (ID), Coffin-Siris syndrome (CSS), autism spectrum disorders (ASD),and schizophrenia (SCZ) (Celen et al., 2017). Further, Jung et al. (2017) reported that the heterozygous deletion of Arid1b led to the disruption of GABAergic IN development. The main reasons for the impaired development of the cortical INs are the suppression of the proliferative rate of the GE progenitors, as well as the enhancement of the apoptosis of GE progenitors during brain development. However, the migratory streams and the rate of the migration showed no change (Jung et al., 2017). In the present research, we uncovered that removal of BAF155 and BAF170 chromatin remodelling subunits dysregulates the development of the cortical IN lineage. Our data endow a novel function of the subunits in orchestrating the differentiation and migration of the subcortical progenitors. At first glance, we observed that the number of the GABAergic INs were reduced in the developing cortex of the BAF155/170 dcKO_Olig2-Cre mutants. It is significant to identify the extent of this alteration in the cortical development has been provoked from the absence of the BAF complex. Consistent with previous *in vivo* works indicated the important function of the Nkx2.1 in directing the INs subpopulation toward their final territory, for example, the migrating INs toward cortex quickly dwindle the Nkx2.1 expression (Villar-Cervino et al., 2015), whereas the MGE derived striatal INs maintains Nkx2.1 expression post-mitotically (Nóbrega-Pereira et al., 2008; Villar-Cervino et al., 2015). The molecular characterization of this mouse model revealed that the number of the Nkx2.1+Sox6+ neurons were decreased in the GP, while other sets of Nkx2.1+ cells were found to accumulate around the mutant MGE/POA region. Thus, these outputs reflect the failure of Nkx2.1 cells to migrate out of the mutant MGEs. On the other hand, we examined the expression of Sox6, which is a key regulator in the differentiation and migration of the dorsal INs (Batista-Brito et al., 2009). Additionally, prior Sox6 cKO mutant studies imparted that Sox6 implicated in the maturation of the MGE emerged IN subtypes particularly somatostatin INs (SST) and parvalbumin INs (PV), but dispensable of proliferation and apoptosis of the MGE progenitor cells (Batista-Brito et al., 2009; Hu et al., 2017; Liodis et al., 2007). Foremost, we detected a significant reduction of the Sox6 expressing INs in the d/vTel at late embryonic age (E18.5). Thus, we speculated that the differentiation and/or the migration of the cortical GABAergic lineage seemed to be disrupted by the loss of BAF complex in the MGEs. For a comprehensive explanation, we examined the expression of Lhx6 TF that acts upstream of Sox6 (Batista-Brito et al., 2009) and downstream of Nkx2.1 (Azim et al., 2009), at mid and late developing stages (E15.5 and E18.5). Of greater significance, Nkx2.1 transcriptional regulator mediates the specification fate of the GABAergic and the cholinergic neural cell lineages through the activation of the Lhx6 and Lhx8 molecular factors (Hu et al., 2017; Liodis et al., 2007; Sandberg et al., 2016; Sussel et al., 1999). Nevertheless, Lhx6 shows significant contribution in cortical INs maturity via induction of cascade genes, involving Cxcr7 and Arx (Sandberg et al., 2016; Vogt et al., 2014). On the other hand, Lhx8 promotes the cholinergic cell lineage development (Fragkouli et al., 2009; Sandberg et al., 2016; Zhao et al., 2003), whereas, Lhx6 and Lhx8 have commonly cooperated in the differentiation of GP and sonic hedgehog (Shh)-derived production of the pallial INs (Flandin et al., 2011; Sandberg et al., 2016). Lhx6 cKO phenotype studies revealed that Lhx6 is insignificant for the MGE born INs specification, although its activity is pivotal for the differentiation of MGE-derived INs to specific IN subpopulations (PV and SSTINs) (Liodis et al., 2007). Additionally, it is indispensable for normal tangential migration of the MGE-born INs to the embryonic pallium (Liodis et al., 2007). Our BAF155/170 −/− mutant analyses revealed an accumulation of the differentiated Lhx6+ cells in the mutant pallidum, consequently, a little number of the Lhx6+ expressing INs were migrated out the MGE/POA region and no Lhx6+ cells were found in the cortex at E15.5. This phenotypic pattern developed even more evidently at an advanced stage (E18.5), most of the Lhx6+ INs was accumulated within dcKO pallidum, while few numbers of Lhx6+ differentiated cells succeeded to migrate along with the dcKO striatum and piriform cortex toward the cerebral cortex. Accordingly, BAF deficit subpallium disrupts potentially the differentiation of the MGE-derived cells, provoking delay and/or misrouting of the GABAergic IN line to travel to the pallium via the piriform cortex. Overall, the conditional deletion of the BAF complex from the developing ventral NSCs resulted in defect of the MGE neurogenesis and abnormal migration of the cortical GABAergic INs.

4.7.   Prospective role of BAF complex in cell fate decision

Cell fate choice is a critical and a multistep process, which is responsible mainly for confining the potency of the progenitors by eliciting the expression of TFs that guide a target cell lineage program. Consistency to the concept of so-called "master regulator" (Ninkovic et al., 2013), a signal of a sole gene in competent cells can elicit the expression of the cascade genes for a specific cell line program, ensuing in an ultimate cell fate commitment (Baker, 2001; Ninkovic et al., 2013). It is unclear whether these key regulators dictate several genes prompting lineage determination, and/ or rule the downstream effectors (Hobert, 2011; Ninkovic et al., 2013). Yet, how these chief regulators act in concert with chromatin-remodelling is unknown. Various master regulators of neuronal lineage fate have been determined in embryonic and postnatal brains, for examples Pax6, SoxC, Mash1, and Dlx2 (Bergsland et al., 2011; Bergsland et al., 2006; Hack et al., 2005; Haubst et al., 2004; Heins et al., 2002; Mu et al., 2012; Ninkovic et al., 2013; Petryniak et al., 2007; Schuurmans et al., 2004). Moreover, previous reports provided evidences, demonstrating the decisive function of Dlx family in promotion of cascade genes that specifically affect neurogenic lines, involving Lhx6, Lhx8, Arx, Mafb, c-Maf, Sp8 and Sp9 (Lindtner et al., 2019). Pai et al. (2019) reported that Mafb and c-Maf are indispensable for MGE derived INs production; however, they work antagonistically in the mature INs to regulate their physiological properties. Further studies elucidated that Sp8 and Sp9 have necessary roles in the differentiation programs of projections neurons in the basal forebrain (Xu et al., 2018). Furthermore, Dlx family suppress neuronal differentiation inhibitors; for instance Id2 and Id4 (Lindtner et al., 2019). Among Dlx family, Dlx2 and Dlx1 are expressed in the GEs (VZ and SVZ) to enhance the pro-neuronal mechanisms (Lindtner et al., 2019; Yun et al., 2002). These programs implicate the inhibition of the genes that regulate the regional fate of the forebrain progenitors, including Olig2, GSX2, Hes5, and Pax6 (Lindtner et al., 2019). Strikingly, the dynamic interplay of Pax6 and Brg1-dependent chromatin remodelling complex, loss of Pax6 or Brg1 *in vitro* and *in vivo* led to similar fate conversion phenotypes from neuronal to glial lineages (Ninkovic et al., 2013). Here, we focused on examining the functional role of BAF complex in presiding fate cell decision via the master activators Dlx1/2. Dlx1 and Dlx2 enhance the GABAergic neurogenic fate at the expense of OL lineage through the inhibition of the TF Olig2 (Sultan and Shi, 2018) and Olig1 (Silbereis et al., 2014). Similarly to our phenotype analysis of BAF null mutants; Dlx1/2 cKO animals unveil a stark reduction in the cortical INs, approximately 70% (Sultan and Shi, 2018) caused by disruption of the differentiation and absence of the migration of GABAergic INs in the cerebral cortex (Le et al., 2017). Nonetheless, our findings revealed expansion of the subpallial MGE (VZ/ SVZ), which is delineated by enhanced Dlx1/2 expression at late embryonic age. Given the remarkable defects in neurogenesis after deletion of Olig1 highlights the critical role of Olig1 in inhibiting the development of cortical INs. Consequently, we speculate that the absence of BAF complex caused the inhibition of the oligodendrocytic lineage supported by the augmentation of Dlx1/2+ germinative cells in the mutant MGEs. Additionally, we also hypothesized that Dlx1/2 master activators were unable to promote the downstream enhancers that granted the proper differentiation and migration of the cortical INs. Although the mechanistic ordinances that regulate the neuron-glial switch are poorly understood, our data raise the potential significance of the BAF 155/170 subunits in guiding the master activator Dlx2 to maintain the neurogenic cell fate versus the glial fate. Further investigations should be undertaken needed to fully recognise the implication of the chromatin remodelling BAF complex in deciphering the neuron-glial fate as a main perspective for the future work.

4.8. Conclusions and future perspectives

BAFopathies, which are set of human neurodevelopmental diseases; involving ASD, ID, SCZ, ALS, Coffin-Siris syndrome (CSS), and Nicolaides-Baraitser syndromes (NCBRS) (Aref-Eshghi et al., 2018; Machol et al., 2019). BAFopathies showed up as a consequence of mutations in genes encoding for the chromatin remodelling BAF subunits (Aref-Eshghi et al., 2018; Machol et al., 2019). Gain- and loss- of function studies have been dedicated to exploring the implication BAF complex in aspects of mammalian brain development (Bischof et al., 2015; Jung et al., 2017; Yu et al., 2013). Nevertheless, its role in the development of GE-derived cells is still not fully understood. Our findings emphasize the pivotal role of the BAF complex in orchestrating the generation of two cell lineages that are emerged from the GE progenitors throughout the embryogenesis. Regarding the OL cell line, our findings indicate an important role of BAF complex in generating healthy OPCs owing typical proliferative capability, and consequently, production of a normal number of the differentiated and myelinated OLs in the developing murine forebrain (Fig.20 A- C). Considering the IN cell line, we demonstrated the substantial function of BAF complex for normal neurogenesis of MGE-born INs, signified by the proper differentiation and migration of INs (Fig.21). However, the mechanisms that mediate the specification of the OL and IN lines are still unclear. Many questions are needed additional investigations in terms of cell fate decision among neuron-glial lines. Indeed, further studies are also required to dedicate the potential association of the BAF complex in controlling the proliferative division capacity of the adult OPCs. Additionally, we propose further investigations to globally elucidate the potentiated function of the BAF complex in the production of the other waves of the cortical INs from the LGE and CGE. Such these works would be necessary for a complete picture of the prospective involvements of BAF complex in governing the production of the OLs and INs in the mammalian telencephalon. These findings endow comprehensive insights for deciphering the potential explanation for certain demyelination diseases, as well as finding prospective mechanisms of some neurodevelopmental disorders caused by the improper development of the cortical INs.

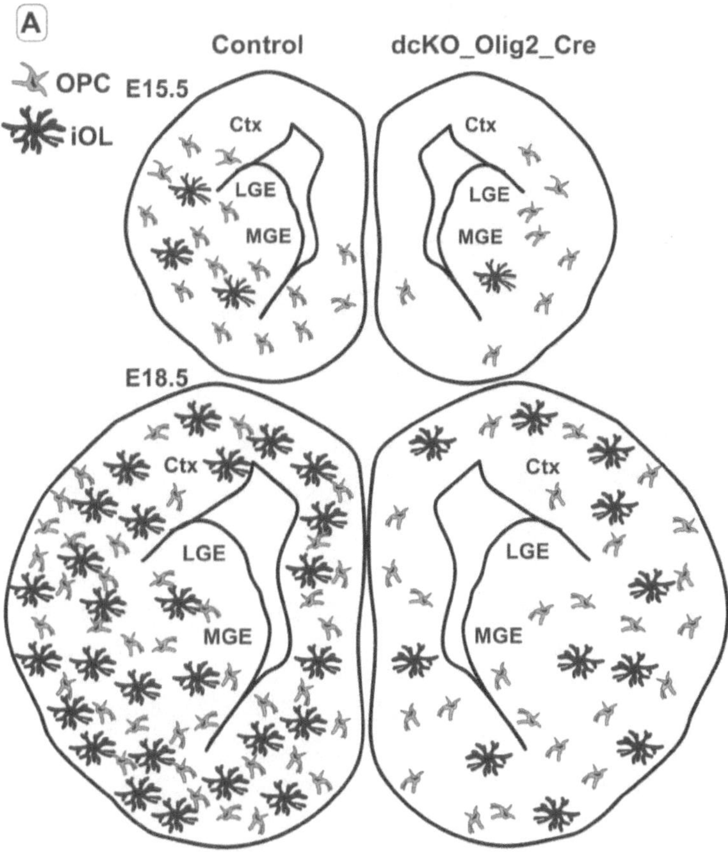

Figure 20 A. Cartoon depicts dcKO_Olig2 Cre_phenotypes in terms of OL (rostral sections).

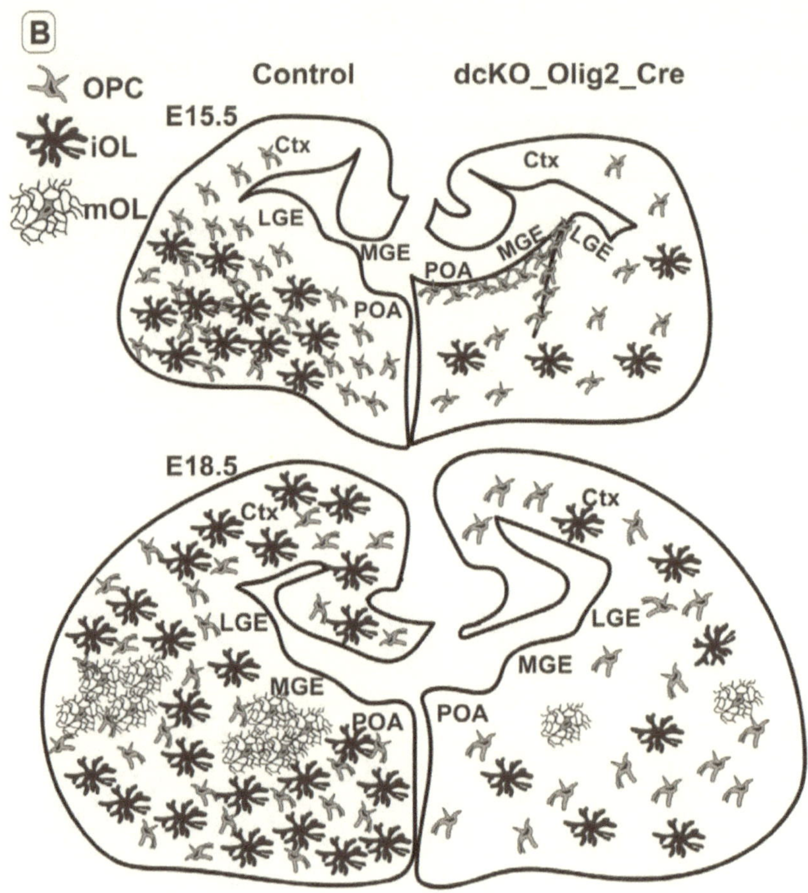

Figure 20 B. Cartoon depicts dcKO_Olig2 Cre_ phenotypes in terms of OL (intermediate sections).

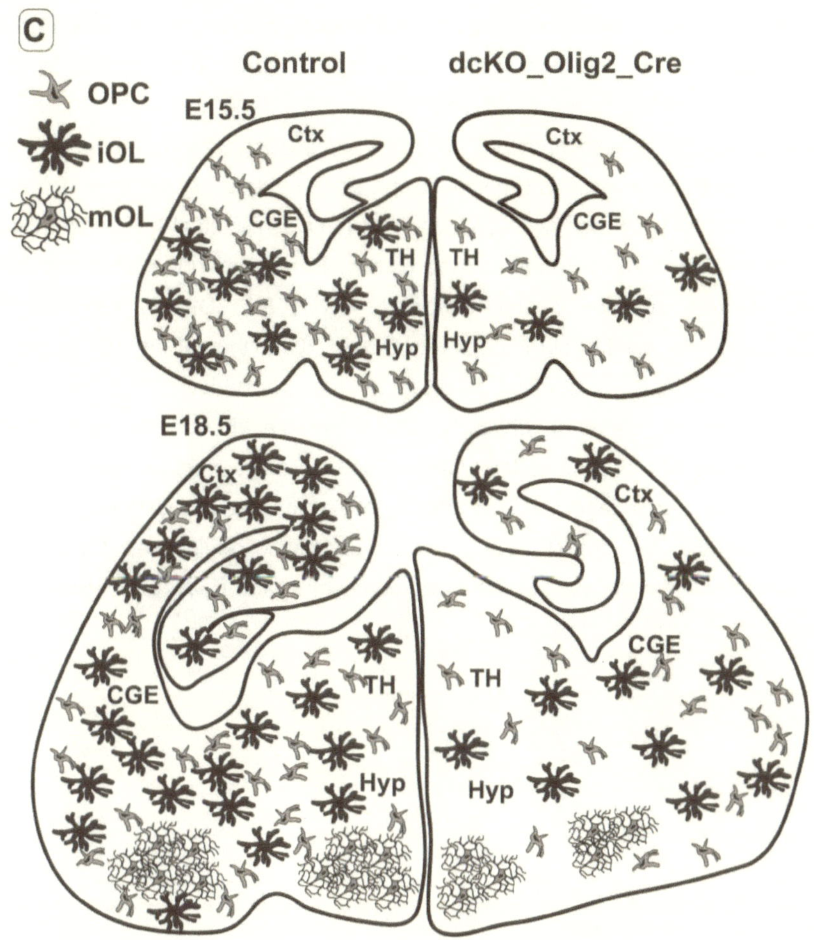

Figure 20 C. Cartoon depicts dcKO_Olig2 Cre_phenotypes in terms of OL (caudal sections).

Figure 20 (A-C). Graphical diagrams illustrate the substantial effects of the conditional loss of BAF complex on proliferation, subsequently differentiation and myelination of the OL production along the embryonic mouse forebrain (derived from this study). (A) Typically, no mOLs have been recorded over the rostral level of the telencephalon; nonetheless, we detected depletion number of iOLs and OPCs in the mutant model compared to control. (B) Normally, few mOLs clusters (PLP+) have started to appear at optic chiasm medial level, nevertheless, we found rarely fewer in number (one or two) in mutants rather than the normal brain. Marked reduction of the number of OPCs and iOLs in subpallium and consequently the number of the migratory OPCs/iOLs to the pallium. (C) Insignificant change in the clusters number of mOLs at the caudal level, however, obvious decrement of the number of OPCs and iOLs in the vTel and accordingly the number of the migrant OPCs to the cortex. OPC, oligodendrocyte precursor cell (PDGFR+); iOL, immature oligodendrocyte (Sox10+); mOL, mature oligodendrocyte (PLP+). Abbreviations: Ctx, cortex; LGE, lateral ganglionic eminence; MGE, medial ganglionic eminence; CGE, caudal ganglionic eminence; POA, pre-optic area; TH, Thalamus; Hyp, hypothalamus.

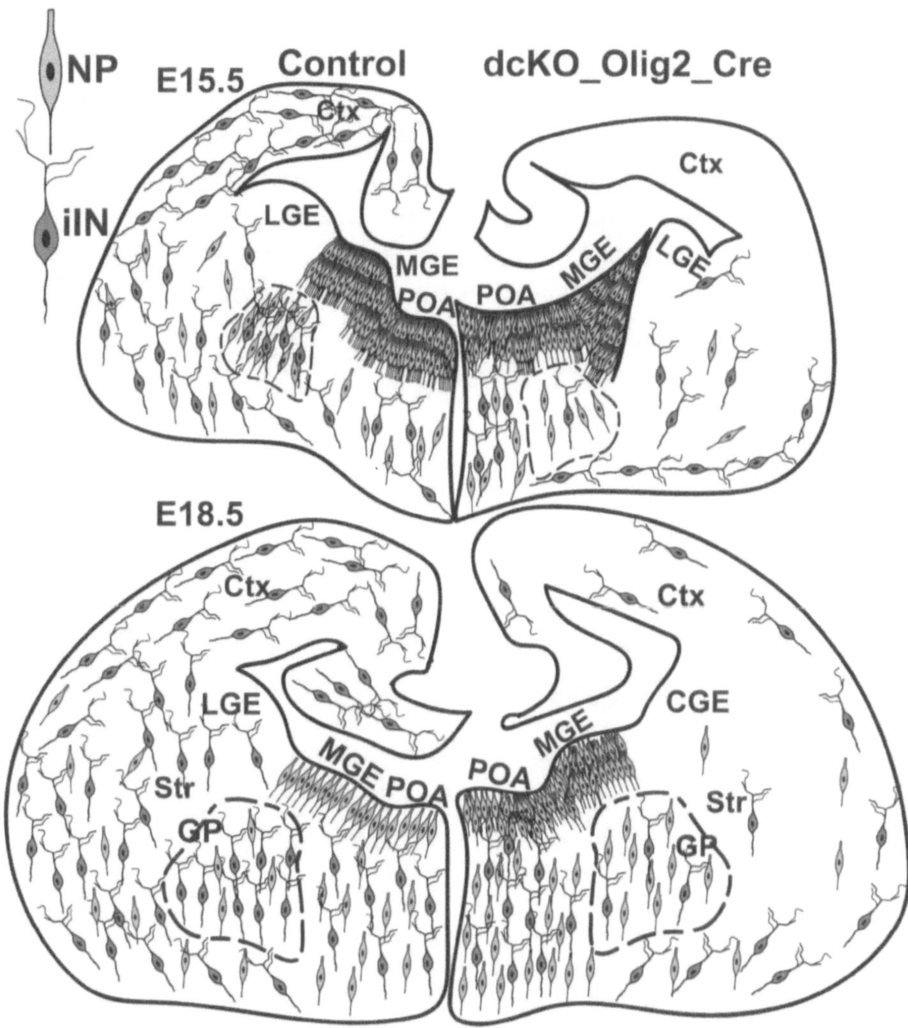

Figure 21. Cartoon depicts aberration in the differentiation and migration of the developing IN caused by the absence of the BAF complex in the developing murine forebrain (intermediate rostrocaudal level) summarized from this study. Normally, at E15.5 migratory routes of INs have been organized to populate the cortex, at the contrary, no cells have been detected along the cortex at this stage. Additionally, significant reduction of NP and iIN in the vTel compared to control. Marked accumulation of NP (Dlx2+/Dlx1+/Nk2.1+) in mutant MGE compared to

control. At E18.5, few number of iIN (Nkx2.1-Lhx6+Sox6+) succeeded to migrate to the pallium, whereas the control brain demonstrates huge number of iIN (Nkx2.1-Lhx6+Sox6+). Moreover, low numbers of NP and iIN have been detected at the striatal area and GP in dcKO_Olig2_Cre mutant model. Expanded mutant MGE was marked compared to control. Abbreviations: NP, neural progenitor (Dlx2+Dlx1; Nkx2.1+Lhx6+Sox6-) ; iIN, immature neuron (Nkx2.1+ Lhx6+Sox6+ OR Nkx2.1-Lhx6+Sox6+); Ctx, cortex; LGE, lateral ganglionic eminence; MGE, medial ganglionic eminence; CGE, caudal ganglionic eminence; POA, pre-optic area; Str, striatum; GP, globus pallidus.

5. Summary

A role for the multi-subunit BAF (mammalian SWI/SNF) complex in brain development has been postulated as its dysfunction causes several neurological diseases, including intellectual disability, schizophrenia, autism spectrum disorders, and amyotrophic lateral sclerosis. Despite this disease-related interest, there is little known about regulatory mechanisms exerted by BAF complex in the proliferative control of the oligodendrocytes (OL) progenitors, differentiation and migration of the GE-derived neuronal progenitors. Herein, we show that BAF complex acts as cell cycle regulator, controlling the dynamic balance between proliferation and differentiation of OL progenitors. OL progenitors are originated from three waves of the ventral and dorsal forebrain. Initially, Nkx2.1 expressing progenitors in the medial ganglionic eminence (MGE) specify into the first wave of oligodendrocyte precursor cells (OPCs) and then they migrate towards the cerebral cortex by E15.5. The second wave of OPCs is produced from the GSX2 expressing progenitors in the lateral and caudal ganglionic eminences (LGE and CGE) and arrives in the pallium at E16.5. Eventually, the third wave of the OPCs is generated after birth from Emx1-expressing cortical progenitors. Committed OPCs divide either symmetrically to produce a pair of cycling OPCs or asymmetrically to give rise to one proliferated OPC and one differentiated OL. Our findings pointed out that conditional loss of BAF complex caused a depleted pool of OL cells in the pallium and subpallium. Remarkably, the deletion of BAF complex impaired the proliferative capacity of the OPCs, and consequently diminished the ability of OLs differentiation and maturation. GE progenitors concurrently produce diverse neuronal and glial cell types. We focused on MGE progenitors since they share an initial developmental birthplace of the most GABAergic interneurons (INs) and the first wave of OPCs. The potential fate of the MGE progenitors is mainly controlled by the master transcription factors Dlx1 and Dlx2, which activate the development of the IN lineage and simultaneously, inhibit the progression of the OL lineage. Furthermore, we demonstrate that BAF complex has a pivotal role in commanding the MGE neurogenesis. Our data revealed that conditional absence of BAF complex in the MGE progenitors lead to severe lessening of the MGE-derived INs, along with impairment of neuronal differentiation and migration. Overall, these findings emphasize that BAF complex is crucial for proper proliferative division and enhance of OPCs and IN differentiation during murine brain development.

5. References

Adams, K.V., and Morshead, C.M. (2018). Neural stem cell heterogeneity in the mammalian forebrain. Progress in neurobiology *170*, 2-36.

Alfert, A., Moreno, N., and Kerl, K. (2019). The BAF complex in development and disease. Epigenetics Chromatin *12*, 19-19.

Alifragis, P., Liapi, A., and Parnavelas, J.G. (2004). Lhx6 regulates the migration of cortical interneurons from the ventral telencephalon but does not specify their GABA phenotype. Journal of Neuroscience *24*, 5643-5648.

Aref-Eshghi, E., Bend, E.G., Hood, R.L., Schenkel, L.C., Carere, D.A., Chakrabarti, R., Nagamani, S.C.S., Cheung, S.W., Campeau, P.M., Prasad, C., *et al.* (2018). BAFopathies' DNA methylation epi-signatures demonstrate diagnostic utility and functional continuum of Coffin–Siris and Nicolaides–Baraitser syndromes. Nature Communications *9*, 4885.

Ascoli, G.A., Alonso-Nanclares, L., Anderson, S.A., Barrionuevo, G., Benavides-Piccione, R., Burkhalter, A., Buzsáki, G., Cauli, B., DeFelipe, J., and Fairén, A. (2008). Petilla terminology: nomenclature of features of GABAergic interneurons of the cerebral cortex. Nature Reviews Neuroscience *9*, 557.

Audia, J.E., and Campbell, R.M. (2016). Histone modifications and cancer. Cold Spring Harbor perspectives in biology *8*, a019521.

Azim, E., Jabaudon, D., Fame, R.M., and Macklis, J.D. (2009). SOX6 controls dorsal progenitor identity and interneuron diversity during neocortical development. Nature neuroscience *12*, 1238.

Baker, N.E. (2001). Master regulatory genes; telling them what to do. Bioessays *23*, 763-766.

Bannister, A.J., and Kouzarides, T. (2011). Regulation of chromatin by histone modifications. Cell research *21*, 381-395.

Batista-Brito, R., Rossignol, E., Hjerling-Leffler, J., Denaxa, M., Wegner, M., Lefebvre, V., Pachnis, V., and Fishell, G. (2009). The cell-intrinsic requirement of Sox6 for cortical interneuron development. Neuron *63*, 466-481.

Bergles, D.E., and Richardson, W.D. (2015). Oligodendrocyte Development and Plasticity. Cold Spring Harbor perspectives in biology *8*, a020453-a020453.

Bergsland, M., Ramsköld, D., Zaouter, C., Klum, S., Sandberg, R., and Muhr, J. (2011). Sequentially acting Sox transcription factors in neural lineage development. Genes & development *25*, 2453-2464.

Bergsland, M., Werme, M., Malewicz, M., Perlmann, T., and Muhr, J. (2006). The establishment of neuronal properties is controlled by Sox4 and Sox11. Genes & development *20*, 3475-3486.

Bischof, M., Weider, M., Küspert, M., Nave, K.-A., and Wegner, M. (2015). Brg1-Dependent Chromatin Remodelling Is Not Essentially Required during Oligodendroglial Differentiation. J Neurosci *35*, 21.

Bond, A.M., Ming, G.-L., and Song, H. (2015). Adult Mammalian Neural Stem Cells and Neurogenesis: Five Decades Later. Cell stem cell *17*, 385-395.

Boshans, L.L., Factor, D.C., Singh, V., Liu, J., Zhao, C., Mandoiu, I., Lu, Q.R., Casaccia, P., Tesar, P.J., and Nishiyama, A. (2019). The Chromatin Environment Around Interneuron Genes in Oligodendrocyte Precursor Cells and Their Potential for Interneuron Reprograming. Frontiers in Neuroscience *13*.

Bradl, M., and Lassmann, H. (2010). Oligodendrocytes: biology and pathology. Acta neuropathologica *119*, 37-53.

Butt, S.J., Fuccillo, M., Nery, S., Noctor, S., Kriegstein, A., Corbin, J.G., and Fishell, G. (2005). The temporal and spatial origins of cortical interneurons predict their physiological subtype. Neuron *48*, 591-604.

Celen, C., Chuang, J.-C., Luo, X., Nijem, N., Walker, A.K., Chen, F., Zhang, S., Chung, A.S., Nguyen, L.H., and Nassour, I. (2017). Arid1b haploinsufficient mice reveal neuropsychiatric phenotypes and reversible causes of growth impairment. Elife *6*, e25730.

Chen, Y.-J.J., Friedman, B.A., Ha, C., Durinck, S., Liu, J., Rubenstein, J.L., Seshagiri, S., and Modrusan, Z. (2017). Single-cell RNA sequencing identifies distinct mouse medial ganglionic eminence cell types. Scientific reports *7*, 45656-45656.

Choi, J., Ko, M., Jeon, S., Jeon, Y., Park, K., Lee, C., Lee, H., and Seong, R.H. (2012). The SWI/SNF-like BAF complex is essential for early B cell development. The Journal of Immunology *188*, 3791-3803.

Cong Tuoc, T., Narayanan, R., and Stoykova, A. (2013). BAF chromatin remodeling complex: cortical size regulation and beyond. Cell cycle *12*, 2953-2959.

Copray, S., Huynh, J., Sher, F., Casaccia, P., and Boddeke, E. (2009). Epigenetic Mechanisms Facilitating Oligodendrocyte Development, Maturation, and Aging. Glia *57*, 1579-1587.

Corbin, J.G., Nery, S., and Fishell, G. (2001). Telencephalic cells take a tangent: non-radial migration in the mammalian forebrain. Nat Neurosci *4 Suppl*, 1177-1182.

Davidoff, M., and Schulze, W. (1990). Combination of the peroxidase anti-peroxidase (PAP)-and avidin-biotin-peroxidase complex (ABC)-techniques: an amplification alternative in immunocytochemical staining. Histochemistry *93*, 531-536.

Dimou, L., Simon, C., Kirchhoff, F., Takebayashi, H., and Götz, M. (2008). Progeny of Olig2-Expressing Progenitors in the Gray and White Matter of the Adult Mouse Cerebral Cortex. J Neurosci *28*, 10434.

Djebali, S., Davis, C.A., Merkel, A., Dobin, A., Lassmann, T., Mortazavi, A., Tanzer, A., Lagarde, J., Lin, W., and Schlesinger, F. (2012). Landscape of transcription in human cells. Nature *489*, 101-108.

Emery, B. (2010). Regulation of oligodendrocyte differentiation and myelination. Science *330*, 779-782.

Emery, B., and Lu, Q.R. (2015). Transcriptional and Epigenetic Regulation of Oligodendrocyte Development and Myelination in the Central Nervous System. Cold Spring Harbor perspectives in biology *7*, a020461-a020461.

Finzsch, M., Stolt, C.C., Lommes, P., and Wegner, M. (2008). Sox9 and Sox10 influence survival and migration of oligodendrocyte precursors in the spinal cord by regulating PDGF receptor alpha expression. Development *135*, 637-646.

Flames, N., Pla, R., Gelman, D.M., Rubenstein, J.L., Puelles, L., and Marín, O. (2007). Delineation of multiple subpallial progenitor domains by the combinatorial expression of transcriptional codes. Journal of Neuroscience *27*, 9682-9695.

Flandin, P., Kimura, S., and Rubenstein, J.L. (2010). The progenitor zone of the ventral medial ganglionic eminence requires Nkx2-1 to generate most of the globus pallidus but few neocortical interneurons. Journal of Neuroscience *30*, 2812-2823.

Flandin, P., Zhao, Y., Vogt, D., Jeong, J., Long, J., Potter, G., Westphal, H., and Rubenstein, J.L. (2011). Lhx6 and Lhx8 coordinately induce neuronal expression of Shh that controls the generation of interneuron progenitors. Neuron *70*, 939-950.

Fragkouli, A., van Wijk, N.V., Lopes, R., Kessaris, N., and Pachnis, V. (2009). LIM homeodomain transcription factor-dependent specification of bipotential MGE progenitors into cholinergic and GABAergic striatal interneurons. Development *136*, 3841-3851.

Furusho, M., Ono, K., Takebayashi, H., Masahira, N., Kagawa, T., Ikeda, K., and Ikenaka, K. (2006). Involvement of the Olig2 transcription factor in cholinergic neuron development of the basal forebrain. Developmental biology *293*, 348-357.

Galloway, D.A., and Moore, C.S. (2016). miRNAs As Emerging Regulators of Oligodendrocyte Development and Differentiation. Frontiers in Cell and Developmental Biology *4*.

García-López, M., Abellán, A., Legaz, I., Rubenstein, J.L., Puelles, L., and Medina, L. (2008). Histogenetic compartments of the mouse centromedial and extended amygdala based on gene expression patterns during development. Journal of Comparative Neurology *506*, 46-74.

Gelman, D., Griveau, A., Dehorter, N., Teissier, A., Varela, C., Pla, R., Pierani, A., and Marin, O. (2011). A wide diversity of cortical GABAergic interneurons derives from the embryonic preoptic area. J Neurosci *31*, 16570-16580.

Gelman, D.M., Martini, F.J., Nóbrega-Pereira, S., Pierani, A., Kessaris, N., and Marín, O. (2009). The Embryonic Preoptic Area Is a Novel Source of Cortical GABAergic Interneurons. J Neurosci *29*, 9380-9389.

Goldman, S.A., and Kuypers, N.J. (2015). How to make an oligodendrocyte. Development *142*, 3983-3995.

Gregath, A., and Lu, Q.R. (2018). Epigenetic modifications—insight into oligodendrocyte lineage progression, regeneration, and disease. FEBS letters *592*, 1063-1078.

Hack, M.A., Saghatelyan, A., de Chevigny, A., Pfeifer, A., Ashery-Padan, R., Lledo, P.-M., and Götz, M. (2005). Neuronal fate determinants of adult olfactory bulb neurogenesis. Nature neuroscience *8*, 865-872.

Halder, R., Hennion, M., Vidal, R.O., Shomroni, O., Rahman, R.-U., Rajput, A., Centeno, T.P., Van Bebber, F., Capece, V., and Vizcaino, J.C.G. (2016). DNA methylation changes in plasticity genes accompany the formation and maintenance of memory. Nature neuroscience *19*, 102.

Hardy, R.J., and Friedrich, V.L. (1996). Oligodendrocyte progenitors are generated throughout the embryonic mouse brain, but differentiate in restricted foci. Development *122*, 2059.

Hargreaves, D.C., and Crabtree, G.R. (2011). ATP-dependent chromatin remodeling: genetics, genomics and mechanisms. Cell Research *21*, 396-420.

Haubst, N., Berger, J., Radjendirane, V., Graw, J., Favor, J., Saunders, G.F., Stoykova, A., and Götz, M. (2004). Molecular dissection of Pax6 function: the specific roles of the paired domain and homeodomain in brain development. Development *131*, 6131-6140.

Heins, N., Malatesta, P., Cecconi, F., Nakafuku, M., Tucker, K.L., Hack, M.A., Chapouton, P., Barde, Y.-A., and Götz, M. (2002). Glial cells generate neurons: the role of the transcription factor Pax6. Nature neuroscience *5*, 308-315.

Hensch, T.K. (2005). Critical period plasticity in local cortical circuits. Nature Reviews Neuroscience *6*, 877.

Hernández-Miranda, L.R., Parnavelas, J.G., and Chiara, F. (2010). Molecules and mechanisms involved in the generation and migration of cortical interneurons. ASN neuro *2*, e00031-e00031.

Ho, L., and Crabtree, G.R. (2010). Chromatin remodelling during development. Nature *463*, 474-484.

Hobert, O. (2011). Regulation of terminal differentiation programs in the nervous system. Annual review of cell and developmental biology *27*, 681-696.

Hu, J.S., Vogt, D., Lindtner, S., Sandberg, M., Silberberg, S.N., and Rubenstein, J.L. (2017). Coup-TF1 and Coup-TF2 control subtype and laminar identity of MGE-derived neocortical interneurons. Development *144*, 2837-2851.

Inder, T.E., and Huppi, P.S. (2000). In vivo studies of brain development by magnetic resonance techniques. Mental retardation and developmental disabilities research reviews *6*, 59-67.

Jakovcevski, I., Filipovic, R., Mo, Z., Rakic, S., and Zecevic, N. (2009). Oligodendrocyte development and the onset of myelination in the human fetal brain. Frontiers in neuroanatomy *3*, 5.

Jakovcevski, I., and Zecevic, N. (2005). Sequence of oligodendrocyte development in the human fetal telencephalon. Glia *49*, 480-491.

Jin, Z., and Liu, Y. (2018). DNA methylation in human diseases. Genes & diseases *5*, 1-8.

Jung, E.M., Moffat, J.J., Liu, J., Dravid, S.M., Gurumurthy, C.B., and Kim, W.Y. (2017). Arid1b haploinsufficiency disrupts cortical interneuron development and mouse behavior. Nat Neurosci *20*, 1694-1707.

Kadoch, C., and Crabtree, G.R. (2015). Mammalian SWI/SNF chromatin remodeling complexes and cancer: Mechanistic insights gained from human genomics. Science advances *1*, e1500447.

Kanatani, S., Yozu, M., Tabata, H., and Nakajima, K. (2008). COUP-TFII Is Preferentially Expressed in the Caudal Ganglionic Eminence and Is Involved in the Caudal Migratory Stream. J Neurosci *28*, 13582-13591.

Kelsom, C., and Lu, W. (2013). Development and specification of GABAergic cortical interneurons. Cell & Bioscience *3*, 19.

Kessaris, N., Fogarty, M., Iannarelli, P., Grist, M., Wegner, M., and Richardson, W.D. (2006). Competing waves of oligodendrocytes in the forebrain and postnatal elimination of an embryonic lineage. Nature neuroscience *9*, 173.

Kessaris, N., Magno, L., Rubin, A.N., and Oliveira, M.G. (2014). Genetic programs controlling cortical interneuron fate. Current opinion in neurobiology *26*, 79-87.

Kidder, B.L., Palmer, S., and Knott, J.G. (2009). SWI/SNF-Brg1 regulates self-renewal and occupies core pluripotency-related genes in embryonic stem cells. Stem cells *27*, 317-328.

Lavdas, A.A., Grigoriou, M., Pachnis, V., and Parnavelas, J.G. (1999). The medial ganglionic eminence gives rise to a population of early neurons in the developing cerebral cortex. Journal of Neuroscience *19*, 7881-7888.

Le, T.N., Zhou, Q.-P., Cobos, I., Zhang, S., Zagozewski, J., Japoni, S., Vriend, J., Parkinson, T., Du, G., and Rubenstein, J.L. (2017). GABAergic interneuron differentiation in the basal forebrain is mediated through direct regulation of glutamic acid decarboxylase isoforms by Dlx homeobox transcription factors. Journal of Neuroscience *37*, 8816-8829.

Li, H., He, Y., Richardson, W.D., and Casaccia, P. (2009). Two-tier transcriptional control of oligodendrocyte differentiation. Current opinion in neurobiology *19*, 479-485.

Lim, L., Mi, D., Llorca, A., and Marín, O. (2018). Development and functional diversification of cortical interneurons. Neuron *100*, 294-313.

Lindtner, S., Catta-Preta, R., Tian, H., Su-Feher, L., Price, J.D., Dickel, D.E., Greiner, V., Silberberg, S.N., McKinsey, G.L., and McManus, M.T. (2019). Genomic Resolution of DLX-Orchestrated Transcriptional Circuits Driving Development of Forebrain GABAergic Neurons. Cell reports *28*, 2048-2063. e2048.

Liodis, P., Denaxa, M., Grigoriou, M., Akufo-Addo, C., Yanagawa, Y., and Pachnis, V. (2007). Lhx6 activity is required for the normal migration and specification of cortical interneuron subtypes. J Neurosci *27*, 3078-3089.

Liu, Z., Hu, X., Cai, J., Liu, B., Peng, X., Wegner, M., and Qiu, M. (2007). Induction of oligodendrocyte differentiation by Olig2 and Sox10: Evidence for reciprocal interactions and dosage-dependent mechanisms. Developmental biology *302*, 683-693.

Love, M.I., Huber, W., and Anders, S. (2014). Moderated estimation of fold change and dispersion for RNA-seq data with DESeq2. Genome biology *15*, 550.

Lu, R., and Barca, O. (2012). Fine-Tuning Oligodendrocyte Development by microRNAs. Frontiers in Neuroscience *6*.

Lunden, J.W., Durens, M., Phillips, A.W., and Nestor, M.W. (2019). Cortical interneuron function in autism spectrum condition. Pediatr Res *85*, 146-154.

Ma, D.K., Bonaguidi, M.A., Ming, G.-L., and Song, H. (2009). Adult neural stem cells in the mammalian central nervous system. Cell research *19*, 672-682.

Machol, K., Rousseau, J., Ehresmann, S., Garcia, T., Nguyen, T.T.M., Spillmann, R.C., Sullivan, J.A., Shashi, V., Jiang, Y.-h., and Stong, N. (2019). Expanding the spectrum of BAF-related disorders: de novo variants in SMARCC2 cause a syndrome with intellectual disability and developmental delay. The American Journal of Human Genetics *104*, 164-178.

Madisen, L., Zwingman, T.A., Sunkin, S.M., Oh, S.W., Zariwala, H.A., Gu, H., Ng, L.L., Palmiter, R.D., Hawrylycz, M.J., and Jones, A.R. (2010). A robust and high-throughput Cre reporting and characterization system for the whole mouse brain. Nature neuroscience *13*, 133.

Maki, T., Liang, A., Miyamoto, N., Lo, E., and Arai, K. (2013). Mechanisms of oligodendrocyte regeneration from ventricular-subventricular zone-derived progenitor cells in white matter diseases. Frontiers in Cellular Neuroscience *7*.

Marín, O. (2012). Interneuron dysfunction in psychiatric disorders. Nature Reviews Neuroscience *13*, 107.

Marın, O., Anderson, S.A., and Rubenstein, J.L. (2000). Origin and molecular specification of striatal interneurons. Journal of Neuroscience *20*, 6063-6076.

Marín, O., and Rubenstein, J.L. (2001). A long, remarkable journey: tangential migration in the telencephalon. Nature Reviews Neuroscience *2*, 780-790.

Masahira, N., Takebayashi, H., Ono, K., Watanabe, K., Ding, L., Furusho, M., Ogawa, Y., Nabeshima, Y.-i., Alvarez-Buylla, A., and Shimizu, K. (2006). Olig2-positive progenitors in the embryonic spinal cord give rise not only to motoneurons and oligodendrocytes, but also to a subset of astrocytes and ependymal cells. Developmental biology *293*, 358-369.

Mashtalir, N., D'Avino, A.R., Michel, B.C., Luo, J., Pan, J., Otto, J.E., Zullow, H.J., McKenzie, Z.M., Kubiak, R.L., St Pierre, R., *et al.* (2018). Modular Organization and Assembly of SWI/SNF Family Chromatin Remodeling Complexes. Cell *175*, 1272-1288 e1220.

Matsumoto, S., Banine, F., Feistel, K., Foster, S., Xing, R., Struve, J., and Sherman, L.S. (2016). Brg1 directly regulates Olig2 transcription and is required for oligodendrocyte progenitor cell specification. Developmental biology *413*, 173-187.

Medina, L., and Abellán, A. (2012). The mouse nervous system.

Minocha, S., Valloton, D., Arsenijevic, Y., Cardinaux, J.-R., Guidi, R., Hornung, J.-P., and Lebrand, C. (2017). Nkx2. 1 regulates the generation of telencephalic astrocytes during embryonic development. Scientific reports *7*, 1-20.

Miyoshi, G., Butt, S.J., Takebayashi, H., and Fishell, G. (2007). Physiologically distinct temporal cohorts of cortical interneurons arise from telencephalic Olig2-expressing precursors. Journal of Neuroscience *27*, 7786-7798.

Moore, L.D., Le, T., and Fan, G. (2013). DNA methylation and its basic function. Neuropsychopharmacology *38*, 23-38.

Mu, L., Berti, L., Masserdotti, G., Covic, M., Michaelidis, T.M., Doberauer, K., Merz, K., Rehfeld, F., Haslinger, A., and Wegner, M. (2012). SoxC transcription factors are required for neuronal differentiation in adult hippocampal neurogenesis. Journal of Neuroscience *32*, 3067-3080.

Nakatani, H., Martin, E., Hassani, H., Clavairoly, A., Maire, C.L., Viadieu, A., Kerninon, C., Delmasure, A., Frah, M., and Weber, M. (2013). Ascl1/Mash1 promotes brain oligodendrogenesis during myelination and remyelination. Journal of Neuroscience *33*, 9752-9768.

Narayanan, R., Pirouz, M., Kerimoglu, C., Pham, L., Wagener, R.J., Kiszka, K.A., Rosenbusch, J., Seong, R.H., Kessel, M., and Fischer, A. (2015). Loss of BAF (mSWI/SNF) complexes causes global transcriptional and chromatin state changes in forebrain development. Cell reports *13*, 1842-1854.

Narayanan, R., and Tuoc, T.C. (2014). Roles of chromatin remodeling BAF complex in neural differentiation and reprogramming. Cell and tissue research *356*, 575-584.

Naruse, M., Ishizaki, Y., Ikenaka, K., Tanaka, A., and Hitoshi, S. (2017). Origin of oligodendrocytes in mammalian forebrains: a revised perspective. The Journal of Physiological Sciences *67*, 63-70.

Nave, K.-A. (2010). Myelination and support of axonal integrity by glia. Nature *468*, 244-252.

Newville, J., Jantzie, L.L., and Cunningham, L.A. (2017). Embracing oligodendrocyte diversity in the context of perinatal injury. Neural Regen Res *12*, 1575-1585.

Nguyen, H., Kerimoglu, C., Pirouz, M., Pham, L., Kiszka, K.A., Sokpor, G., Sakib, M.S., Rosenbusch, J., Teichmann, U., and Seong, R.H. (2018). Epigenetic regulation by BAF complexes limits neural stem cell proliferation by suppressing Wnt signaling in late embryonic development. Stem cell reports *10*, 1734-1750.

Ninkovic, J., Steiner-Mezzadri, A., Jawerka, M., Akinci, U., Masserdotti, G., Petricca, S., Fischer, J., von Holst, A., Beckers, J., and Lie, C.D. (2013). Essential role of BAF complex interacting with Pax6 in establishment of a core cross-regulatory neurogenic network. Cell stem cell *13*, 403.

Niquille, M., Limoni, G., Markopoulos, F., Cadilhac, C., Prados, J., Holtmaat, A., and Dayer, A. (2018). Neurogliaform cortical interneurons derive from cells in the preoptic area. Elife *7*, e32017.

Nóbrega-Pereira, S., Kessaris, N., Du, T., Kimura, S., Anderson, S.A., and Marín, O. (2008). Postmitotic Nkx2-1 controls the migration of telencephalic interneurons by direct repression of guidance receptors. Neuron *59*, 733-745.

Ohtomo, R., Iwata, A., and Arai, K. (2018). Molecular mechanisms of oligodendrocyte regeneration in white matter-related diseases. International journal of molecular sciences *19*, 1743.

Ono, K., Takebayashi, H., Ikeda, K., Furusho, M., Nishizawa, T., Watanabe, K., and Ikenaka, K. (2008). Regional-and temporal-dependent changes in the differentiation of Olig2 progenitors in the forebrain, and the impact on astrocyte development in the dorsal pallium. Developmental biology *320*, 456-468.

Pai, E.L.-L., Vogt, D., Clemente-Perez, A., McKinsey, G.L., Cho, F.S., Hu, J.S., Wimer, M., Paul, A., Darbandi, S.F., and Pla, R. (2019). Mafb and c-Maf have prenatal compensatory and postnatal antagonistic roles in cortical interneuron fate and function. Cell reports *26*, 1157-1173. e1155.

Petryniak, M.A., Potter, G.B., Rowitch, D.H., and Rubenstein, J.L. (2007). Dlx1 and Dlx2 control neuronal versus oligodendroglial cell fate acquisition in the developing forebrain. Neuron *55*, 417-433.

Phelan, M.L., Sif, S., Narlikar, G.J., and Kingston, R.E. (1999). Reconstitution of a core chromatin remodeling complex from SWI/SNF subunits. Molecular cell *3*, 247-253.

Poitras, L., Ghanem, N., Hatch, G., and Ekker, M. (2007). The proneural determinant MASH1 regulates forebrain <em>Dlx1/2</em> expression through the I12b intergenic enhancer. Development *134*, 1755.

Puelles, L., Harrison, M., Paxinos, G., and Watson, C. (2013). A developmental ontology for the mammalian brain based on the prosomeric model. Trends in neurosciences *36*, 570-578.

Puelles, L., Kuwana, E., Puelles, E., Bulfone, A., Shimamura, K., Keleher, J., Smiga, S., and Rubenstein, J.L. (2000). Pallial and subpallial derivatives in the embryonic chick and mouse telencephalon, traced by the expression of the genes Dlx-2, Emx-1, Nkx-2.1, Pax-6, and Tbr-1. Journal of Comparative Neurology *424*, 409-438.

Puelles, L., Morales-Delgado, N., Merchán, P., Castro-Robles, B., Martínez-de-la-Torre, M., Díaz, C., and Ferran, J.L. (2016). Radial and tangential migration of telencephalic somatostatin neurons originated from the mouse diagonal area. Brain Structure and Function *221*, 3027-3065.

Que, L., Winterer, J., and Földy, C. (2019). Deep survey of GABAergic interneurons: emerging insights from gene-isoform transcriptomics. Frontiers in molecular neuroscience *12*, 115.

Richardson, W.D., Kessaris, N., and Pringle, N. (2006). Oligodendrocyte wars. Nature Reviews Neuroscience *7*, 11-18.

Sandberg, M., Flandin, P., Silberberg, S., Su-Feher, L., Price, J.D., Hu, J.S., Kim, C., Visel, A., Nord, A.S., and Rubenstein, J.L. (2016). Transcriptional networks controlled by NKX2-1 in the development of forebrain GABAergic neurons. Neuron *91*, 1260-1275.

Schüller, U., Heine, V.M., Mao, J., Kho, A.T., Dillon, A.K., Han, Y.-G., Huillard, E., Sun, T., Ligon, A.H., and Qian, Y. (2008). Acquisition of granule neuron precursor identity is a critical determinant of progenitor cell competence to form Shh-induced medulloblastoma. Cancer cell *14*, 123-134.

Schuurmans, C., Armant, O., Nieto, M., Stenman, J.M., Britz, O., Klenin, N., Brown, C., Langevin, L.M., Seibt, J., and Tang, H. (2004). Sequential phases of cortical specification involve Neurogenin-dependent and-independent pathways. The EMBO journal *23*, 2892-2902.

Selten, M., van Bokhoven, H., and Kasri, N.N. (2018). Inhibitory control of the excitatory/inhibitory balance in psychiatric disorders. F1000Research *7*.

Silbereis, John C., Nobuta, H., Tsai, H.-H., Heine, Vivi M., McKinsey, Gabriel L., Meijer, Dimphna H., Howard, MacKenzie A., Petryniak, Magda A., Potter, Gregory B., Alberta, John A., *et al.* (2014). Olig1 Function Is Required to Repress Dlx1/2 and Interneuron Production in Mammalian Brain. Neuron *81*, 574-587.

Sokpor, G., Xie, Y., Rosenbusch, J., and Tuoc, T. (2017). Chromatin Remodeling BAF (SWI/SNF) Complexes in Neural Development and Disorders. Frontiers in molecular neuroscience *10*, 243.

Somogyi, P., Freund, T., Wu, J.-Y., and Smith, A. (1983). The section-Golgi impregnation procedure. 2. Immunocytochemical demonstration of glutamate decar□ylase in Golgi-impregnated neurons and in their afferent synaptic boutons in the visual cortex of the cat. Neuroscience *9*, 475-490.

Spassky, N., Goujet-Zalc, C., Parmantier, E., Olivier, C., Martinez, S., Ivanova, A., Ikenaka, K., Macklin, W., Cerruti, I., Zalc, B., *et al.* (1998). Multiple Restricted Origin of Oligodendrocytes. J Neurosci *18*, 8331.

Staiger, J.F., Möck, M., Proenneke, A., and Witte, M. (2015). What types of neocortical GABAergic neurons do really exist? e-Neuroforum *6*, 49-56.

Stolt, C.C., Rehberg, S., Ader, M., Lommes, P., Riethmacher, D., Schachner, M., Bartsch, U., and Wegner, M. (2002). Terminal differentiation of myelin-forming oligodendrocytes depends on the transcription factor Sox10. Genes Dev *16*, 165-170.

Sugiarto, S., Persson, A.I., Munoz, E.G., Waldhuber, M., Lamagna, C., Andor, N., Hanecker, P., Ayers-Ringler, J., Phillips, J., and Siu, J. (2011). Asymmetry-defective oligodendrocyte progenitors are glioma precursors. Cancer cell *20*, 328-340.

Sultan, K.T., and Shi, S.-H. (2018). Generation of diverse cortical inhibitory interneurons. Wiley interdisciplinary reviews Developmental biology 7, 10.1002/wdev.1306.

Sussel, L., Marin, O., Kimura, S., and Rubenstein, J. (1999). Loss of Nkx2. 1 homeobox gene function results in a ventral to dorsal molecular respecification within the basal telencephalon: evidence for a transformation of the pallidum into the striatum. Development *126*, 3359-3370.

Tamamaki, N., Sugimoto, Y., Tanaka, K., and Takauji, R. (1999). Cell migration from the ganglionic eminence to the neocortex investigated by labeling nuclei with UV irradiation via a fiber-optic cable. Neuroscience research *35*, 241-251.

Touzot, A., Ruiz-Reig, N., Vitalis, T., and Studer, M. (2016). Molecular control of two novel migratory paths for CGE-derived interneurons in the developing mouse brain. Development *143*, 1753-1765.

Tripodi, M., Filosa, A., Armentano, M., and Studer, M. (2004). The COUP-TF nuclear receptors regulate cell migration in the mammalian basal forebrain. Development *131*, 6119-6129.

Tuoc, T., Dere, E., Radyushkin, K., Pham, L., Nguyen, H., Tonchev, A.B., Sun, G., Ronnenberg, A., Shi, Y., Staiger, J.F., *et al.* (2017). Ablation of BAF170 in Developing and Postnatal Dentate Gyrus Affects Neural Stem Cell Proliferation, Differentiation, and Learning. Mol Neurobiol *54*, 4618-4635.

Tuoc, T.C., Boretius, S., Sansom, S.N., Pitulescu, M.E., Frahm, J., Livesey, F.J., and Stoykova, A. (2013). Chromatin regulation by BAF170 controls cerebral cortical size and thickness. Dev Cell *25*, 256-269.

Tuoc, T.C., and Stoykova, A. (2008). Trim11 modulates the function of neurogenic transcription factor Pax6 through ubiquitin-proteosome system. Genes Dev *22*, 1972-1986.

Turrero Garcia, M., and Harwell, C.C. (2017). Radial glia in the ventral telencephalon. FEBS Lett *591*, 3942-3959.

van Tilborg, E., de Theije, C.G.M., van Hal, M., Wagenaar, N., de Vries, L.S., Benders, M.J., Rowitch, D.H., and Nijboer, C.H. (2018). Origin and dynamics of oligodendrocytes in the developing brain: Implications for perinatal white matter injury. Glia *66*, 221-238.

Villar-Cervino, V., Kappeler, C., Nóbrega-Pereira, S., Henkemeyer, M., Rago, L., Nieto, M.A., and Marín, O. (2015). Molecular mechanisms controlling the migration of striatal interneurons. Journal of Neuroscience *35*, 8718-8729.

Vogt, D., Hunt, R.F., Mandal, S., Sandberg, M., Silberberg, S.N., Nagasawa, T., Yang, Z., Baraban, S.C., and Rubenstein, J.L. (2014). Lhx6 directly regulates Arx and CXCR7 to determine cortical interneuron fate and laminar position. Neuron *82*, 350-364.

Wagener, R.J., Dávid, C., Zhao, S., Haas, C.A., and Staiger, J.F. (2010). The somatosensory cortex of reeler mutant mice shows absent layering but intact formation and behavioral activation of columnar somatotopic maps. Journal of Neuroscience *30*, 15700-15709.

Wamsley, B., and Fishell, G. (2017). Genetic and activity-dependent mechanisms underlying interneuron diversity. Nature Reviews Neuroscience *18*, 299.

Wang, J., Duncan, D., Shi, Z., and Zhang, B. (2013). WEB-based gene set analysis toolkit (WebGestalt): update 2013. Nucleic acids research *41*, W77-W83.

Watson, C., Paxinos, G., and Puelles, L. (2012). The Mouse Nervous System.

Wichterle, H., Garcia-Verdugo, J.M., Herrera, D.G., and Alvarez-Buylla, A. (1999). Young neurons from medial ganglionic eminence disperse in adult and embryonic brain. Nature neuroscience *2*, 461-466.

Wichterle, H., Turnbull, D.H., Nery, S., Fishell, G., and Alvarez-Buylla, A. (2001). In utero fate mapping reveals distinct migratory pathways and fates of neurons born in the mammalian basal forebrain. Development *128*, 3759-3771.

Wilson, S.W., and Rubenstein, J.L. (2000). Induction and dorsoventral patterning of the telencephalon. Neuron *28*, 641-651.

Wu, J.I., Lessard, J., and Crabtree, G.R. (2009). Understanding the words of chromatin regulation. Cell *136*, 200-206.

Xu, Q., Tam, M., and Anderson, S.A. (2008). Fate mapping Nkx2.1-lineage cells in the mouse telencephalon. Journal of Comparative Neurology *506*, 16-29.

Xu, W., Lakshman, N., and Morshead, C.M. (2017). Building a central nervous system: The neural stem cell lineage revealed. Neurogenesis (Austin, Tex) *4*, e1300037-e1300037.

Xu, Z., Liang, Q., Song, X., Zhang, Z., Lindtner, S., Li, Z., Wen, Y., Liu, G., Guo, T., and Qi, D. (2018). SP8 and SP9 coordinately promote D2-type medium spiny neuron production by activating Six3 expression. Development *145*.

Yoo, A.S., and Crabtree, G.R. (2009). ATP-dependent chromatin remodeling in neural development. Current opinion in neurobiology *19*, 120-126.

Yu, Y., Chen, Y., Kim, B., Wang, H., Zhao, C., He, X., Liu, L., Liu, W., Wu, L.M., Mao, M., *et al.* (2013). Olig2 targets chromatin remodelers to enhancers to initiate oligodendrocyte differentiation. Cell *152*, 248-261.

Yun, K., Fischman, S., Johnson, J., de Angelis, M.H., Weinmaster, G., and Rubenstein, J.L. (2002). Modulation of the notch signaling by Mash1 and Dlx1/2 regulates sequential specification and differentiation of progenitor cell types in the subcortical telencephalon. Development *129*, 5029-5040.

Yun, K., Potter, S., and Rubenstein, J.L. (2001). Gsh2 and Pax6 play complementary roles in dorsoventral patterning of the mammalian telencephalon. Development *128*, 193-205.

Zhao, Y., Marín, O., Hermesz, E., Powell, A., Flames, N., Palkovits, M., Rubenstein, J.L., and Westphal, H. (2003). The LIM-homeobox gene Lhx8 is required for the development of many cholinergic neurons in the mouse forebrain. Proceedings of the National Academy of Sciences *100*, 9005-9010.

Zhu, X., Hill, R.A., Dietrich, D., Komitova, M., Suzuki, R., and Nishiyama, A. (2011). Age-dependent fate and lineage restriction of single NG2 cells. Development *138*, 745-753.

Zhuo, L., Theis, M., Alvarez-Maya, I., Brenner, M., Willecke, K., and Messing, A. (2001). hGFAP-cre transgenic mice for manipulation of glial and neuronal function in vivo. genesis *31*, 85-94.

Zintzsch, A. (2013). Herausforderungen bei der Belastungsbeurteilung genetisch veränderter Tiere: Wie können Phänotypisierungsdaten und der Blick in die EU helfen? Leipziger Blaue Hefte, 436.